Vegetables and Fruits

THE TIME-LIFE ENCYCLOPEDIA OF GARDENING

Vegetables and Fruits

by

JAMES UNDERWOOD CROCKETT

and

the Editors of TIME-LIFE BOOKS

Watercolor Illustrations by
Richard Crist

An Owl Book
Henry Holt and Company
NEW YORK

THE TIME-LIFE ENCYCLOPEDIA OF GARDENING

EDITORIAL STAFF FOR VEGETABLES AND FRUITS:
EDITOR: Ogden Tanner
Picture Editor: Grace Brynolson
Designer: Leonard Wolfe
Staff Writers: Helen Barer,
Marian Gordon Goldman, Gerry Schremp
Researchers: Sandra Streepey, Gail Cruikshank, Bea Hsia,
Catherine Ireys, Nancy Jacobsen, Jill McManus,
Vivian Stephens
Design Assistant: Anne B. Landry

Library of Congress Cataloging in Publication Data
Crockett, James Underwood.
Vegetables and fruits.
(The Time-Life encyclopedia of gardening)
Reprint. Originally published: New York : Time-Life
Books, 1972.
Bibliography: p.
Includes index.
1. Vegetable gardening. 2. Fruit-culture. 3. Herb
gardening. I. Time-Life Books. II. Title. III. Series.
[SB321.C825 1986]   635   85-27326
ISBN 0-03-008527-6 (pbk.)

First published by Time-Life Books in 1972.
First Owl Book Edition—1986
Printed in the United States of America
10 9 8 7 6 5 4 3 2 1

ISBN 0-03-008527-6

AUTHOR AND CONSULTANT JAMES UNDERWOOD CROCKETT: Author of 13 of the volumes in the Encyclopedia, co-author of two additional volumes, and consultant on other books in the series, James Underwood Crockett has been a lover of the earth and its good things since his boyhood on a Massachusetts fruit farm. He was graduated from the Stockbridge School of Agriculture at the University of Massachusetts and has worked ever since in horticulture. A perennial contributor to leading gardening magazines, he also writes a monthly bulletin, *Flowery Talks,* that is widely distributed through retail florists. His television program, "Crockett's Victory Garden," shown all over the United States, is constantly winning new converts to the Crockett approach to growing things.

THE ILLUSTRATOR: Richard Crist, who provided the 98 watercolor paintings of vegetables, fruits, nuts and herbs on pages 78-147, was trained at Carnegie Institute of Technology and The Art Institute of Chicago. His portrayals of native American wildflowers have appeared in *Natural History* magazine. An amateur gardener and botanist, he has written and illustrated several children's books.

GENERAL CONSULTANTS: Norman F. Childers, Rutgers University, New Brunswick, N.J. Melvin Hoffman, Cornell University, Ithaca, N.Y. Albert P. Nordheden, New York City. Adelma Grenier Simmons, North Coventry, Conn. Robert E. Young, University of Massachusetts, Suburban Experiment Station, Waltham. Staff of the Brooklyn Botanic Garden: Elizabeth Scholtz, Acting Director; Robert S. Tomson, Assistant Director; Thomas R. Hofmann, Plant Propagator; George A. Kalmbacher, Plant Taxonomist; Edmund O. Moulin, Horticulturist; Frank Okamura, Gardener.

THE COVER: A glistening harvest of vegetables, home-grown by photographer Dick Meek, includes such favorites as corn, carrots, tomatoes, squash, onions, lettuce, green peas, radishes, beets, sweet peppers and beans.

CONTENTS

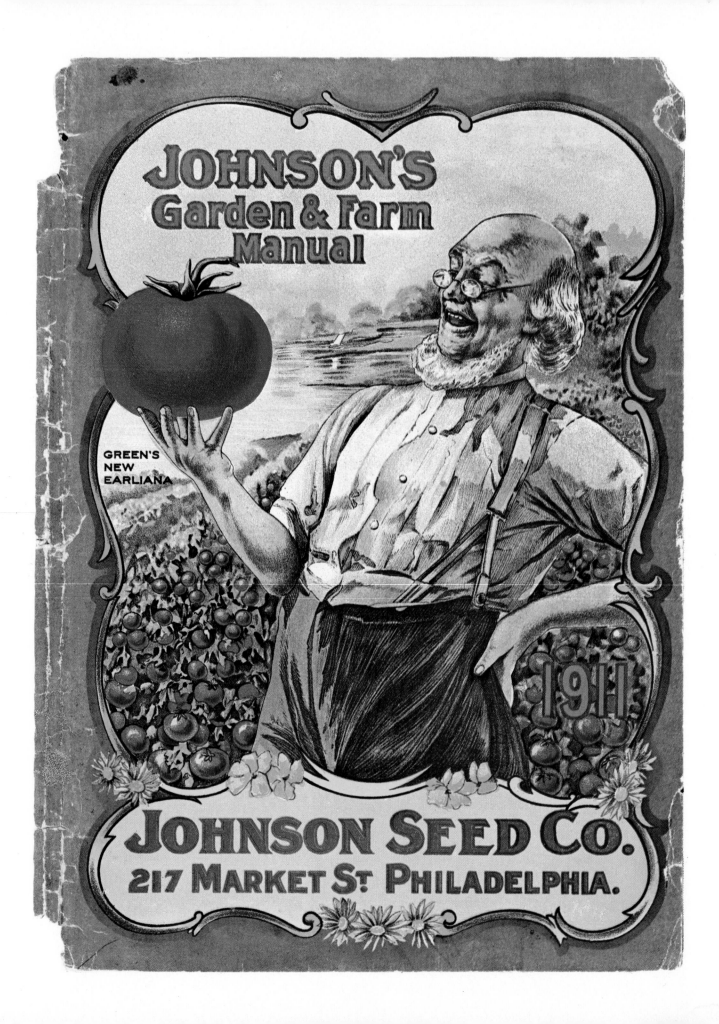

JOHNSON'S
Garden & Farm Manual

GREEN'S
NEW
EARLIANA

1911

JOHNSON SEED CO.
217 MARKET St. PHILADELPHIA.

The joys of growing your own 1

Why, people sometimes ask me, do you grow your own food? Why go to all the trouble of tilling, planting and weeding a piece of your valuable backyard for vegetables? Aren't all those fruit trees and berrybushes a lot of work? Well, I generally answer such questions with a few of my own. When was the last time you bit into a really delicious peach, the juice fairly bursting through the skin? When was the last time you sat down to a steaming plate of fresh asparagus—the tender just-ripe tips, not the stringy kind you generally get at the supermarket? When was the last time you could even find sweet corn picked fresh enough to be really sweet, or raspberries plump enough to make cold cereal a gourmet dish?

The answer, of course, is pretty much what I expected: all too long ago. For even if most Americans are well fed, most of them are also missing the incomparable taste of truly fresh vegetables and fruits. Those small, often hard tomatoes you buy at the store were probably picked when they were still green; moreover, they are of a variety developed not so much for their flavor as for their shipping and handling qualities—including a tough protective skin. Chances are, particularly if you buy them during the off season, that they came from a big mechanized farm in a distant state—California, for example, produces a quarter of all the table food sold in the United States, including nearly two thirds of the tomatoes. Actually the produce industry supplies us with an incredible harvest —refrigerated, frozen, canned, dried—and much of it available almost everywhere at almost every season of the year. What the industry cannot do at long range and in large quantities, however, is to supply most vegetables and fruits the way people like to think of them: truly fresh. To get them that way, you either have to live next to a friendly truck gardener—or grow them yourself.

If you grow your own, you certainly get freshness and you may also save some money—but then again you may not. On the face of it you can beat store prices every time, but when you stop to add up the cost of fertilizers, peat moss, mulch, pesticides, garden

A proud gardener contemplates the fruits of his labor—an eye-popping tomato—on the cover of a 1911 seed catalogue. Earliana varieties are still grown, but disease-resistant types have all but replaced them.

hose, tools, and other odds and ends (not even counting your own labor), you will be lucky to break even. No matter. You will enjoy several things money cannot buy, including the unique satisfaction of growing precisely what you want and eating it at its prime. You can experience at first hand the age-old miracle of tiny seeds becoming bountifully producing plants. And, too, a little spading, hoeing and weed pulling never hurt anyone's waistline.

SUMMER-LONG BOUNTY

My wife, Margaret, and I have had a vegetable and fruit garden for some 30 years, and we would not be without it. It is about 60 by 100 feet, a bit larger than most homeowners might care to keep up, but we find that it is worth every minute we invest in it, and we do not feel obliged to plant every square inch of it every year. We grow a wide assortment of crops for our own enjoyment and to give to friends, and we find ourselves looking forward to each one as it comes into season, actually planning meals around it. In May we feast on asparagus and rhubarb, as well as on the first early plantings of leaf lettuce and radishes. June is strawberry month, with big juicy berries for breakfast, strawberry shortcake for family dinners and—a special treat—Margaret's own strawberry-rhubarb pie *(right)*. July brings carrots and beets as well as tender green peas to go with New England's traditional Fourth of July salmon. It also brings potatoes; they take up a fair amount of space but we always grow some, not so much for the mature potatoes, which are generally no better than those we buy at the market (potatoes ship and keep relatively well), but for the new potatoes, dug when they are only about an inch or two across and possess a delicate flavor that is lost soon after harvesting. August brings peppers, eggplants, tomatoes, sweet corn and summer squashes, and we are picking broccoli and Brussels sprouts well into fall. These are our stand-bys, but every year we try to plant at least a half row of a new vegetable or a new variety of an old one, just to experiment. One year it was an All-America Selection, Waltham Butternut squash; the next it was a new seedless watermelon.

We look forward to our annual calendar of fruit every bit as much. Following the strawberries, which we grow in our vegetable garden, we have raspberries, blackberries and blueberries from a dozen or more bushes; pears, plums, apples and walnuts; and grapes from vines of eight different varieties. We also keep a little herb garden going indoors or out year round to supply parsley, rosemary, thyme, dill and other seasonings for our meals.

SOLVING SPACE PROBLEMS

A 60-by-100-foot vegetable garden and all those bushes, trees and herbs still may sound to you like a great deal of space, time and trouble, particularly if you live on a small suburban lot. But one of the beauties of food gardening today is that you can do a surprising amount of it in very little space—even in 15 by 20 feet, the size of the minigarden shown on page 19. In such a garden you can grow enough fresh vegetables and fruits to spice the usual supermarket fare, and get the especially tasty varieties that do not ship well and therefore rarely if every appear in stores; such varieties, and the regions in which they can be grown, are described in the encyclopedia section of this book, beginning on page 79.

You can get still more out of limited gardening space by using some standard techniques, plus a few special tricks:

• Grow successive crops on the same patch of soil, or even mix together different crops that will not interfere with each other, doubling or tripling the total yield (*pages 20 and 21*).

• Grow some crops vertically. Tomatoes and rambling vine crops such as cucumbers and squashes normally take up a lot of ground, but you can save space—and get earlier-maturing, more easily harvested vegetables—by training them up stakes, fences or netting (*drawings, pages 29 and 30*). Small melons can be handled this way; when the individual fruits become too heavy for the vine to support them, you can hold them up in little slings or cradles of string netting or coarse cheesecloth until they are ready to pick.

• Use midget varieties of vegetables. Midget corn takes up less space than standard varieties; it produces smaller ears but they are just as sweet. Even a city dweller can grow cherry tomatoes or the newer and larger Pixie hybrids in a pot or tub on a balcony, rooftop or window sill. An added advantage of all these small varieties is that they take weeks less to mature than their larger brethren; you can have corn or tomatoes earlier in the year, and even more important, the shorter growing season allows Northern gardeners to get big harvests before early fall frosts.

• Grow midget fruit trees, too. Probably the most spectacular examples of plant miniaturization are the so-called dwarf versions of standard apple, peach and pear trees. Not only do they bear full-sized fruit, but they do so at an earlier age than standard trees, and because they grow only 6 to 8 feet high they are much easier to prune, spray and pick. To save even more space, you can use dwarfs in patio tubs or planter boxes; they will grow even smaller because the containers constrict their roots. Or you can train them by espaliering (*page 55*) to grow flat against a fence, trellis or wall.

Tubbed or espaliered plants are only two of the many ways that fruits and vegetables, and herbs for that matter, can be made decorative assets in the home landscape as well as sources of fresh food, doing double or even triple duty for the space they occupy. Grapevines can be trained along a boundary fence to screen the

STRAWBERRY-RHUBARB PIE

A tart-sweet pie that has been a special favorite of New Englanders for generations is a combination of an early-ripening vegetable, rhubarb, and strawberries, the season's first fruit. The author's family recipe calls for 3 cups of firm young rhubarb stems (discard the leaves, which are poisonous). The stems, cut into 1-inch pieces, are mixed with 1 cup of sugar, ¼ teaspoon of salt, ¼ teaspoon of nutmeg, 2 tablespoons of quick-cooking tapioca and ¼ cup of orange juice. These ingredients are put in a 9-inch pie pan lined with pie crust, topped with 1 cup of fresh sliced strawberries, dotted with 1 tablespoon of butter and crisscrossed with pastry strips. The pie is baked at 425° for 40 minutes and served piping hot with whipped cream for topping.

THE EDIBLE LANDSCAPE

yard with attractive foliage in summer and to yield luscious bunch-es of grapes in fall; or they can be grown on an arbor to provide welcome shade in summer, accented by the clusters of ripening fruit overhead. Fast-growing red raspberries or blackberries not only produce delicious fruit but if strategically located make attractive hedges whose dense growth and thorns provide effective barriers. High-bush blueberries can perform the same function, and offer, in turn, pretty white flowers, lovely green leaves, quarts of berries, and a blaze of orange and scarlet foliage in fall. Apple, peach or plum trees can be used in a line along a boundary as a garden backdrop, property marker or privacy screen. The smaller varieties of fruit trees recommended for home gardens do not cast much useful shade, but nut trees such as Chinese chestnuts, pecans and walnuts can be planted as large shade trees as well as producers of nuts.

Out of habit, I suppose, most people dismiss vegetables and herbs to a strictly utilitarian role, hiding them off somewhere behind the kitchen or garage and using "ornamental" plants to dress up the garden. Sophisticated gardeners I know, however, are not afraid to mix some of the more decorative vegetables right in with their flowers in the main garden—which, incidentally, is an excellent way of experimenting with vegetables without building a separate garden for them. A few clumps of rhubarb, for example, can be grown in a perennial border; the tall fernlike foliage of asparagus looks lovely at the back of a flower border long after the early-growing sprouts have been picked for the table. The red stems and leaf veins of some varieties of Swiss chard make a bright accent in a garden of annuals. Instead of relegating strawberries to the vegetable patch, I have seen homeowners plant them as an

(continued on page 14)

Truck garden on a rooftop

Stewart Mott's home had no land—he lived many stories aboveground in a Manhattan apartment house—but every year he raised eggplants, beans, lettuce, tomatoes, corn and other vegetables for his table. His small truck garden grew in pots and boxes on the roof of his building.

Mott's method is useful even for suburbanites with small yards or stony soil. All you need is a sunny spot and containers of any size or type as long as they have holes in the bottom to let excess water run out. Fill them with a 1-inch drainage layer of coarse gravel and a lightweight growing mixture: either packaged potting soil or a mix of equal parts of vermiculite and peat moss, with 1 tablespoon each of limestone and a slow-release high-phosphorus fertilizer per gallon pailful of mix. Your vegetables may need more frequent watering than those grown in the ground, but otherwise they are just as easy to raise—and easier to reach.

Pot-grown tomatoes (right), along with eggplants, peppers and cabbages (left), get a head start in a cold frame in Stewart Mott's rooftop garden.

Chives arch gracefully from a window box where they can be readily snipped for use in soups or salads, outside Mott's dining room. Most herbs are very easy to grow in containers, either outdoors or indoors on a sunlit kitchen window sill.

Lima beans climb a trellis in the foreground of this pastoral city scene. At left are tubs of corn; beyond them is a tripod entwined with cucumbers. In the center background is a young walnut tree; pumpkin vines cascade out from its base.

The billowing foliage of Brussels sprouts (right), in front of a row of celery, grows out of a green-painted metal planting box. Started in the cold frame shown on page 11, the sprouts will mature and be ready for harvesting by the first fall frost.

attractive ground cover on a sunny slope in front of the house, and they are downright handsome in a strawberry jar on a patio. Herbs, usually thought of as workaday seasonings, make fine potted plants, borders, and even entire show gardens in themselves *(pages 68-71)*. I know people who actually grow corn in tubs on their front steps, and though they have to hand-pollinate the ears because there are not enough other corn plants in the vicinity to ensure cross-pollination, the stalks make a graceful harvesttime welcome at the door. Even the lowly cabbage, still spurned by some as a workingman's vegetable, is a remarkably beautiful plant, something like a great green rose; I have seen it used imaginatively and effectively in planting beds in more than one fashionable garden.

SOME GARDENING BASICS

Whether you make full use of the decorative values of your vegetables, fruits and herbs or grow them simply for good eating, their health and your gardening success will depend largely on attention to a few basics. The first of these is properly prepared soil. Almost any soil can be made fit for growing if you work into it enough organic matter in the form of decayed leaves, compost, peat moss, sawdust, ground bark or manure. All these add body to light sandy soils, giving them the needed capacity to hold moisture and nutrients; they also loosen heavy clay soils so that air and water can penetrate. My favorite organic matter is compost, a particularly rich blend of decayed material that can easily be made from leaves, grass clippings, weeds and other vegetation.

A compost pile can be built in any number of ways; one of the simplest is to stick three or four 4-foot stakes into the ground about a foot deep, then wrap around them a circular or square enclosure of small-mesh wire fencing 3 feet high. Besides leaves and lawn clippings (do not include the latter if you have recently used a weed killer on your lawn), you can throw in dead plant stems or the cut-off tops of beets, carrots and other vegetables. On top of each 4- to 6-inch layer of material, sprinkle about a pint of 5-10-5 fertilizer and a dusting of ground limestone, then cover with a 2-inch blanket of soil and make the top slightly concave to catch rain. Wet the pile from time to time if rainfall is light, and turn it over every four to six weeks so that material on the outside of the pile becomes incorporated into the brewing middle. The compost will be ready to use when it is dark and crumbly, in about three to six months. A 2-inch application each spring, worked into the soil before planting, will make a garden enormously productive.

ACID AND ALKALINE SOILS

Compost will take care of most soils' shortcomings, but it will not guarantee the right degree of acidity or alkalinity, which is measured on a pH scale that runs from 0, for extremely acid, to 7, for neutral, to 14, for extremely alkaline. Most vegetables, fruits and herbs thrive in soils with a pH of 6.0 to 7.0, but each has its own range, as indicated in the encyclopedia, and some require more acid or more alkaline conditions. If your soil does not fall within the proper range for the plants you are growing, you should correct it by adding the proper materials. You can easily determine the pH of your soil with an inexpensive kit sold at garden centers, or by taking or sending a soil sample to your local agricultural extension service for testing.

To raise the pH of an overly acid soil ½ to 1 unit, add 5 pounds of finely ground limestone to each 100 square feet of planting area. To lower the pH of alkaline soils, use either finely ground sulfur, which is slow acting but relatively long lasting, or iron sulfate or aluminum sulfate, which work quickly but dissipate more rapidly in the soil. One half pound of ground sulfur or 3 pounds of iron or aluminum sulfate per 100 square feet will lower the pH by ½ to 1 unit. For especially heavy soils, increase all these amounts by about one third. Any materials used to change the pH of soil should be incorporated deeply into it when it is dug over, and if possible, should be applied in the fall prior to spring planting or in spring prior to fall planting, to give them time to take effect.

A soil that has been properly prepared with organic matter and adjusted to the optimum acid-alkaline level will give plants a good start, but most will need additional feeding at intervals to replenish the natural minerals in the soil as the plants use them up. This is most easily done with chemical fertilizers, which not only have higher concentrations of mineral nutrients than do organic materials such as compost or manure, but also release them more quickly so the plants can use them. For most vegetables I use a dry bagged fertilizer such as 5-10-5, composed of 5 pounds of nitrogen, 10 pounds of phosphorus and 5 pounds of potassium per 100 pounds (the balance is, in effect, filler material that holds and dilutes the chemicals, allowing them to be spread evenly in the garden without burning the plants' roots). For leafy vegetables such as lettuce, cabbage and spinach, I use a fertilizer with a ratio of 10-10-10 to provide a little more nitrogen, which stimulates leaf growth, and potassium, which builds strong root systems.

Whether these minerals come from a chemical fertilizer or from an organic source, they are exactly the same by the time they have become dissolved in the moisture in the soil, which is the only form in which the plants can assimilate them. I use both organic and chemical growing aids in my own garden. With this combination I get bigger, healthier plants that resist insects and diseases, plants that grow faster and produce larger, tastier crops. And that, of course, is the object of the game.

THE UNPOPULAR TOMATO

Of all the native American vegetables —including beans, corn, peppers, potatoes and squash—the most popular today is the tomato. Yet the tomato took the longest to win acceptance in the United States. A staple of Central and South American Indians, the tomato was first introduced to Europe by the conquistadors, where it was quickly adopted by the Italians and French. North Americans, however, did not welcome their own native vegetable until well into the 19th Century, partly because of the belief that it was poisonous (the leaves and stems are toxic). And as late as 1860 Godey's Lady's Book advised that tomatoes should "always be cooked three hours." Today tomatoes are more widely grown at home than any other vegetable; one seed catalogue offers 31 varieties.

BURPEE'S
WHITE
EVERGREEN

BURPEE'S
GOLDEN
BANTAM

Planning and planting vegetables 2

Some of the neighbors in the town where I live have a standing contest to see who can produce the first ripe tomato each year. From time to time, to keep the competition on its toes, one or another resorts to a devious trick such as buying a few tomatoes at the store, tying them to the vines in the dark of night, and then casually strolling out and "picking" them in full view of a neighbor the next morning. One cool rainy summer, the winner confessed to having used a sun lamp on one of his vines, and another year one gardener's plants were so far ahead that no one could figure out his secret—until someone discovered that he had buried a soil-heating cable in his garden to help things along.

The latter two exploits are hardly standard gardening practices, but they do illustrate two of the most vital ingredients in vegetable gardening: light and warmth, without which most vegetables simply will not grow. The best way to get both, of course, is not from a sun lamp or a heating cable, but from the sun itself.

So in planning any vegetable garden, the first thing you should do is locate it where it will get as much sun as it possibly can for a maximum amount of time each day. On a small suburban lot with houses nearby and an average number of trees, such spots are not always in generous supply, but most people can find at least one small area where the sun reaches for five or six hours between the long shadows of morning and evening. If only part of the garden will get this minimum quota, plan on putting sun-loving vegetables such as tomatoes, corn, cucumbers and melons there; others such as lettuce and pumpkins, which tolerate cooler weather and more shade, can be planted at the end where the shadows linger.

Try to locate the garden away from trees or large shrubs: not only will their shade stunt your vegetables, but their broadly ranging roots will steal most of the moisture and nutrients from your carefully fertilized and watered beds. If you have to put your garden close to trees, make sure they will not block the southerly sun; then, to keep the roots from invading the garden, line a nar-

When these two corn varieties were planted for a 1919 catalogue, white corn was for people, yellow corn for cattle. But the new Golden Bantam, the first tender yellow type, quickly became a favorite vegetable.

row trench 2 to 3 feet deep with sheet aluminum, sheet plastic or heavy tar paper as a vertical barrier; you may have to cut off a few tree roots in the process but as long as they are not too large or too numerous the trees will survive.

If you can find a sunny spot close to the house, so much the better; you will be able to carry your fresh-picked produce directly into the kitchen and you will not have to string multiple lengths of hose out to the garden to provide water on dry days.

As a final step in locating your garden, check any potential site for drainage. A garden set in a low swampy spot will not dry out until late in spring—if it dries out at all. A depression at the bottom of a slope can pose a problem of air drainage as well as water drainage; cold air, being heavier than warm air, runs downhill and collects in such places, making them the spots where frost lingers longest in spring and strikes first in fall. On a level or slightly sloping site it does not make much difference if the vegetable rows are laid out on a north-south or east-west axis, but if the land slopes rather sharply, the rows should follow the contour of the slope rather than run downhill; this will prevent quick erosion from runoff during a rainstorm and also help to retain water in dry weather.

CHOOSING YOUR CROPS

In a sunny well-drained location, you should be able to grow many kinds of vegetables, providing you give them good enough soil to grow in (*page 14*) and ample water and food. But to get the most out of your garden, you would do well to plan it with the following basic guidelines in mind:

• *Start with vegetables that taste notably better picked fresh.* Among home-grown vegetables that are usually superior to the same produce bought at the supermarket are tomatoes, asparagus, sweet corn, peas, beans, young onions and carrots, lettuce and summer squash. Winter squash and mature potatoes and onions, on the other hand, keep well and are generally available in good quality in stores, though you can often improve on these too.

• *Choose varieties that will grow satisfactorily in your climate.* Consult the encyclopedia section of this book for some of the best varieties recommended for various regions. It is also a good idea to talk with your local county agricultural extension agent or state agricultural college; many publish pamphlets for home gardeners that list additional locally preferred varieties and also give detailed advice on plantingtimes and growing practices peculiar to the area.

• *Know your local frost dates.* In most parts of the country, the date of the last expected spring frost in your area as well as the date of the first expected fall frost determines the plantingtimes of many vegetables, which cannot stand freezing weather. These dates are given in general terms on the maps on page 149, but there can be

A minigarden in 15 by 20 feet

Anyone with a little space to spare can grow an ample supply of fresh vegetables in a garden planted as illustrated in the diagram below. This minigarden, designed by the author, has been separated by gravel paths into four beds that are small enough to reach into easily; they are raised to minimize bending and make gardening easier, to provide good drainage, to keep soil, gravel and lawn grass in their places and to provide an attractive pattern both winter and summer. The vegetable beds are enclosed by 2-by-8-inch boards that are treated against rotting with a copper naphthenate stain, fastened with angle irons and sunk into the ground (drawing, right). To make multiple use of the space, fast-growing crops like lettuce and radishes are sown among slower-growing ones like corn and melons and harvested before the latter crowd them; early and later crops are also sown in succession to ensure a continuous yield from the same garden space.

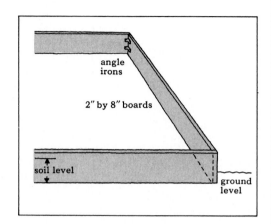

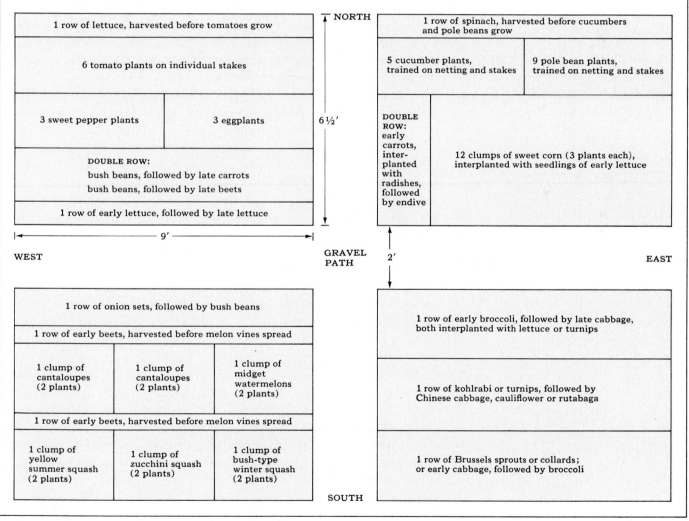

19

wide variations within a given region depending on altitude, nearness to large bodies of water and other factors, and it is best to check with your local weather bureau or county agent for more precise dates. In my own garden in Massachusetts the last spring frost comes about May 15 and the first fall frost arrives about September 25. On a hill only a half mile away, however, where cold air does not accumulate, the last spring frost occurs two weeks earlier, about May 1, and the first fall frost does not come until about October 15. The growing season is more than a month longer on the hill than in the valley.

• *Plant cool-weather crops early or late in the year*. Many crops, especially those of the cabbage family, are so hardy, or cold resistant, that they can be planted as early as the soil can be worked in the spring, usually about six weeks before the last frost date, and they will grow past the fall frost date until the ground freezes. The growing season of other plants, notably spinach and many kinds of peas and lettuce, is limited to cool weather. In Northern areas of the country a crop or two can be grown in the spring and another in the fall, but not in the middle of summer; in the South and Southwest crops can be grown in the winter as well.

• *Plant warm-weather crops when temperatures remain above the required level*. The planting season for crops that need warm weather—melons, tomatoes, eggplants and sweet potatoes, for example —is not reckoned by frost dates, but by the arrival of temperatures that stay above a certain level night and day *(encyclopedia)*. It is futile to plant vegetables such as these when the temperatures are lower and the soil is cold and wet; the seeds will not sprout, or if they should sprout, the seedlings will stand still, in gardener's parlance, and refuse to grow any further until warm weather arrives. You can, however, start these plants indoors or in a cold frame or a hotbed for later planting outdoors.

• *Make successive plantings of one vegetable, or plant early, midseason and late-maturing varieties to distribute the harvest over a longer period*. If you plant all your beans at the same time, you will have to pick them all at about the same time, and even bean lovers have a limit as to how much they can eat at once. It is far better to make one planting after the ground has warmed, then subsequent plantings at two-week intervals until midsummer to assure a moderate but steady supply. In the case of vegetables like tomatoes and cabbages, which are available in varieties that mature at different times, set out plants of early, midseason and late types, if the length of your growing season permits, to spread the production period over two to three months.

• *Know each vegetable's average time to maturity*. The length of time it takes various vegetables to ripen, from planting to maturity, is given in the encyclopedia section. Most seed catalogues and seed packets also give this information in a range of days, for example, "Super-Duper Tomato: 75-80 days." These figures are averages; a northerly climate or a relatively cool cloudy spring and summer can lengthen the ripening period of a 75-day tomato to as much as 90 days, while a southerly climate and ample sun can reduce the period to 70 days. Moreover, the catalogues and seed packets sometimes fail to make it clear that the number of days specified refers to the maturingtime of seeds sown outdoors for most vegetables, but not all; in the case of tomatoes, peppers and eggplants, vegetables that are usually started indoors well ahead of planting, the number of days refers to the time needed to mature after four- to eight-week-old plants have been set out in the garden.

• *Start successive plantings in a nursery bed to save garden space.* The classic way to save space in a home garden is to sow the seeds of another crop as soon as the first one is harvested. But there is an even better way to utilize space and to save several weeks of growingtime, especially in the case of crops like lettuce, cabbage, broccoli and cauliflower, all of which can be grown as second crops following earlier ones. The secret is to sow the seeds in a nursery bed a few weeks before the crop they are to succeed is ready for harvest. Set aside a 3- to 4-foot-square patch of ground or a corner of the garden that is not in use, and scatter seeds of the second, or follow, crop in it; the seeds do not have to be in rows and can be quite close together, since the plants will be moved before they become full-sized. When the seedlings are 2 to 3 inches tall, dig them up with a little soil around their roots and transplant them into the rows where you have harvested the first crop.

• *Grow more plants in a given area by interplanting.* Cabbages, for example, need a spacing of about 1½ to 2 feet; if a young lettuce plant is set between each pair of cabbages, the fast-growing lettuces will be ready to harvest before the slower-growing cabbages overshadow them. Double or triple rows of small vegetables such as lettuce, beets, carrots, spinach and radishes can be planted in bands only 12 to 18 inches wide between walkways, narrow enough to make weeding easy. Onion sets (little onion bulbs) can be planted in very early spring; two months later, when the weather becomes warm enough, tomato plants can be set in the same row, taking the places of young onions pulled for eating as tasty green onions, sometimes called scallions; the rest of the onions will be ready for harvest at maturity before the tomatoes grow big enough to crowd them. Spinach can be grown between rows of tomatoes; by the time the tomato plants are half grown, the spinach will already have been harvested. A row of snap beans can be planted between two rows of parsnips; the beans will have

ALL-AMERICA VEGETABLES

The designation All-America Winner or All-America Selection, often seen in seed catalogues, identifies a vegetable as one of the best all-round choices in its class. The title is given by All-America Selections, a nonprofit organization founded by seedsmen in 1932 to encourage the development of worthy new varieties. Every year scores of entries are tested at 27 trial grounds located in the United States, Canada and Mexico. Superior entries are given merit citations and medals.

been picked and the old plants can be pulled up before the slow-growing parsnips expand to fill the space.

• *Time your plantings so you will be around to enjoy the harvest.* You might be surprised at the number of amateur gardeners who enthusiastically plant a patch of sweet corn, for example, without looking at the number of days marked on the packet, only to find themselves vacationing on a beach far away when the ears come ripe. Even an understanding neighbor may not be able to use your ripened crops while you are away—particularly if he has to pick them himself.

• *Locate perennial vegetables out of the way.* Most vegetables are annuals; for all practical purposes they have a life span of one growing season and must be replanted every year. The two most notable and widely grown exceptions are asparagus and rhubarb, which may continue to produce for as long as 20 to 30 years from one planting. Both asparagus and rhubarb—as well as beds of strawberries and perennial herbs such as sage, thyme and chives—should be set at one side of the garden or in completely separate beds. This is the only way you can be sure you will not inadvertently disturb them when you dig the soil for the annual crops each year.

• *Place tall-growing vegetables so that they will not shade lower-*

PLANTING ASPARAGUS ROOTS

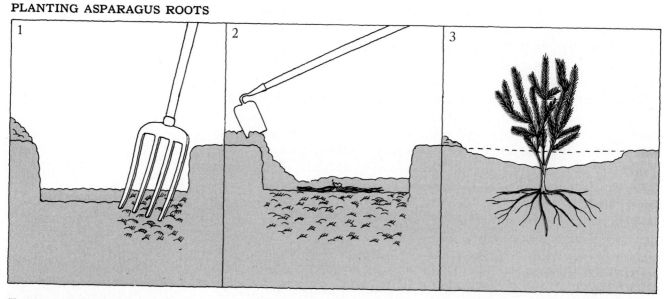

To plant asparagus roots, dig a trench 2 feet wide and 8 inches deep, piling the soil to one side. Work 2 inches of rotted manure and a sprinkling of 5-10-5 fertilizer into the bottom to a depth of 8 inches.

After raking the trench bottom smooth and walking on it to firm it, space plants 18 inches apart, crown side up and roots spread out in all directions. Cover the plants with 2 inches of topsoil from the pile.

As the plants sprout their foliage during the summer, gradually add some of the remaining pile of topsoil around the growing stems. By late summer the trench should be filled, level with the original soil surface.

growing ones. Corn, pole beans and tomatoes or cucumbers grown on stakes or fences should be set at the north side of the garden where they will not block the sun from lower plants.

• *Save your unused seeds for next year—but don't bother trying to salvage seeds from your crops.* Many seed packets contain more seeds of a given vegetable than the small-scale gardener can use in one season. Packets are usually stamped with a "packed for" date, but if properly stored, the seeds of most vegetables can be used successfully from one to six years beyond this date, as noted in the encyclopedia section. Even though seed is relatively cheap, there is no point in throwing it away while it is still usable. When you have opened the packet and have made the first sowing, place the remaining seeds, in the packet, in an airtight container such as a Mason jar and put it on the back shelf of the refrigerator. The packet may be taken out for any subsequent sowings, then returned to storage for use the following year. If there is no room in the refrigerator, store the jar in the coolest place in the house; freezing will not hurt the seeds. There will usually be a slight drop in the germination rate of stored seeds from year to year, but you will still get plenty of plants from them.

While it is possible to save seeds produced by some of your own garden plants, it is generally not very practical. Many plants, such as carrots, beets, parsnips and most members of the cabbage family, are biennials; we never see their flowers or the seeds that follow because they are not produced until the second year, long after garden plants have been eaten. Even in the case of most vegetables that do produce seeds the first season, such as tomatoes, the virtues of extracting and saving the seeds are marginal for the amateur. Most of the better vegetable varieties are hybrids, crosses between two or more types that have been bred to give the result better flavor, texture, size or resistance to disease. Seeds saved from hybrids often do not run true, as horticulturists say; they have a way of reverting to one or another of their ancestors, and while they are perfectly edible they are generally inferior in taste, color, shape or other qualities to the improved strain.

PREPARING THE SOIL

Once you have considered these basic aspects of vegetable gardening, you can get out your tools and go to work. Digging and preparing the garden in fall rather than spring has much to recommend it. Not only does it allow pH balancers like limestone and sulfur the required amount of time to do their work, but in fall the soil is warm, relatively dry and in relatively better condition to cultivate compared with spring, when heavy soils in particular are still full of moisture and can be hopelessly compacted into bricklike clods if trampled on too soon. In addition, there are usually not as many urgent gardening and household chores to be done in fall as there are in spring.

If you have a fair-sized garden, it may pay you to buy or rent a rotary tilling machine. Try to get the kind with the blades mounted behind the motor; it leaves the soil smoother after the tiller has passed. Rotary tilling or hand digging should go down about 8 inches except for such deep-rooted crops as parsnips; for most crops tilling any deeper merely brings less desirable subsoil up into the root zone. Go over the area once with a spading fork or tiller to break up the soil, then rake stones and clods from the surface with an iron-toothed rake. Next, add organic matter—2 inches of compost, peat moss, manure, sawdust or ground bark, or 3 to 4 inches of partly decayed leaves. For every 100 square feet of area scatter 4 to 5 pounds of a balanced chemical fertilizer such as 5-10-5, or 2 to 3 pounds of 10-10-10, just before planting. If you use manure or compost reduce the amount of chemical fertilizer by about one quarter; if you use sawdust or ground bark increase the amount of fertilizer by one quarter to compensate for the nitrogen used up by these materials as they decay. If you use dry bagged manure instead of fresh rotted manure as organic matter, put on a ½-inch layer instead of a 2-inch one because it is more highly concentrated. Till these materials into the soil, cross-

SOWING SEEDS OUTDOORS

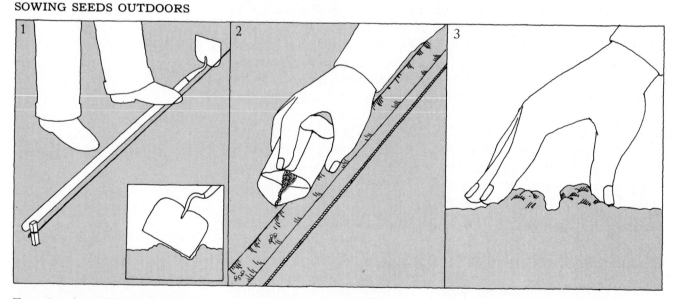

To make a furrow for small seeds lay a hoe beside a guideline of twine and press the handle to the desired depth (encyclopedia). For large seeds like those of beans, cut V-shaped furrows with a hoe (inset).

Sow small seeds by squeezing the open packet lengthwise and tapping lightly to slide the seeds out three or four at a time, spacing groups at the distances recommended. Large seeds are set in place by hand.

To cover most seeds, pinch the furrow and pat it flat. Very small seeds should be covered not with soil but with a thin layer of sifted compost or vermiculite. For large seeds, use a hoe to tamp the soil.

wise to the first pattern, to work them into the top 8 inches. Rake the surface smooth once more and you are ready to plant.

Most vegetables are started by sowing seeds directly in the garden, as shown in the drawings at left. As a general rule, seeds should be set about three times as deep as their diameter. Small seeds such as those of carrots and onions, and ones that are slow to germinate, like parsnips, should be covered with finely sifted compost, vermiculite or fine sand; if ordinary soil is used it can easily form a hard crust after wetting and thus inhibit sprouting. Moisten the seedbed with a very fine spray from a garden hose and keep it moist until the seedlings appear. I like to sow a few radish seeds along with slow-sprouting plants such as carrots because the radishes will be up in a few days and mark the rows; this enables me to cultivate close to the rows and get rid of sprouting weeds without disturbing the seeds.

Even beginning gardeners soon find they can get a one- to two-month jump on the growing season by starting certain late-maturing plants such as tomatoes indoors before setting them in the garden. As they progress, they also discover that peppers, eggplants and cabbages, as well as vine crops such as cucumbers, melons and squashes, can be treated the same way. In fact, in Northern regions with short growing seasons that is the only way these plants can be grown satisfactorily before fall frosts set in. Some of these vegetables can be bought already started, but doing it yourself is more rewarding and easier on the pocketbook as well; a single tomato plant bought from a greenhouse when six weeks old can cost as much as a whole package of tomato seeds, enough to start dozens of plants.

You can start seeds indoors in almost any kind of container from a gardener's flat to a flowerpot, then lift the seedlings out one by one when they are ready and plant them outside. One of the simplest and easiest methods, however, and one that subjects the seedlings to the least transplanting shock, is to start the plants in individual pots made of compressed peat or in expandable peat pellets (drawings, page 26) and set them into the garden at the proper time, pots—or pellets—and all.

In setting the plants out, you can space them by eye at the intervals recommended or use a homemade planting board; mine is a plank 4 feet long and 4 inches wide along one edge of which I have cut V-shaped notches every 6 inches, making every other one deeper to mark intervals of 1 foot. The board not only allows me to plan and plant exactly the number of seedlings I want in a row of given length, but it serves as a knee rest to prevent my weight from making large round holes in the soft soil. I choose a cloudy day for transplanting, or do the job in the evening so the plants can be-

CHAMPION VEGETABLES

Although biggest rarely means tastiest, gardeners for centuries have grown outsized vegetables—mainly to see just how large they could get them. As long ago as 1544, a German botanist wrote of radishes said to weigh 100 pounds; more recently a South Carolina gardener produced 3-pound tomatoes larger than grapefruit and an Ohio man has grown squashes of over 200 pounds. For several years the Men's Garden Clubs of America have held an annual Big Pumpkin Contest limited to gardeners under 18 years of age. In 1968 eleven-year-old Lynn Price and her eight-year-old sister Sharon of Marietta, Georgia, set a record: by prudent watering, fertilizing and removing all but the most promising fruit on the vine, they produced a huge—and virtually inedible— pumpkin weighing 235 pounds.

come acclimated overnight before the sun's full rays hit them. I water the seedlings an hour or two before they are to be transplanted to make sure the cells are filled with moisture; after setting them in holes scooped with a trowel I make a little water-holding basin around each plant. To get each plant off to a fast start, I water it with a cupful of a water-soluble fertilizer diluted to the strength recommended for starting or transplanting plants. Then, to keep away cutworms, which can topple a seedling in a hurry, I put a little collar around the base of the stem. You can use a cylinder of cardboard, but ordinary paper cups are even handier; simply punch or cut out the bottom of the cup and slip it carefully over the foliage, pushing it about an inch down into the soil. (If the top of the plant will not squeeze easily through the cup, cut the

STARTING SEEDS IN PEAT PELLETS

1. *To start seeds for tomatoes, eggplants and peppers indoors in the expanding containers known as peat pellets, set the pellets in a tray and pour water over them; they will rise to six or seven times their original height (inset).*

2. *Use a pencil to make three holes in each pellet (inset), insert a seed in each and pinch the peat to cover it. Place the tray in a warm place (70° to 75°); keep the pellets moist, but not soggy. When the seeds sprout, move them to a cool sunny window.*

3. *After their true leaves form, snip off all but the strongest seedling in each pellet. Transfer plants to a cold frame or put them outside for a few hours daily until the weather becomes warm enough to establish them in the garden.*

4. *Plant the seedlings, pellets and all, and build ½-inch-high watering saucers. Push a paper cup with its bottom removed an inch into the soil around each plant to ward off cutworms. Plant spindly tomato seedlings sideways, burying the lower stem (inset) so it can develop roots.*

side of the cup and wrap it, as you would a collar, around the stem.)

To protect tender seedlings from hot sun, I often cut leafy twigs and stick them in the soil on the south side of the plants; by the time the leaves have fallen off a few days later the plants no longer need the protection. To shelter seedlings of warmth-loving plants like melons and cucumbers from cool spring temperatures, and from insects and driving rain, I cover them with wax-paper caps, available from seed companies and garden centers; they act like little individual greenhouses, giving the plants an extra boost early in the season when nights are apt to be cool.

Although you can start tomato, cucumber and other seedlings in a sunny window in the house, if you leave them indoors too long the seedlings will grow too rapidly and become weak and spindly as a result of reaching toward the one-directional source of light. If they are moved directly to the garden while the nights are still cool, such seedlings—overgrown and weak—may not be able to make the sudden adjustment. The best solution is to allow the seeds to germinate indoors, and as soon as the seedlings have formed one set of true leaves above the early rounded seedling leaves, to move the plants to a miniature greenhouse outdoors where they can gradually acclimate to outdoor conditions.

The simplest such greenhouse is an inexpensive cold frame like the one shown in the drawings on page 28. A friend of mine and I designed the frame and he built it in the equivalent of a Saturday afternoon. It is only 4 by 4 feet, its sides cut from half a standard-sized 4-by-8-foot piece of plywood, yet it can hold more than enough seedlings for a good-sized home garden. The top, made of light lumber, is covered with 4- to 6-mil transparent polyethylene sheeting from the hardware store, not only making it lightweight for handling and storing but also eliminating any broken-glass hazard to small children or pets that might climb on or stumble over it. The wood is treated against rot with a preservative stain containing copper naphthenate. (Do not use preparations containing creosote, mercury compounds or pentachlorophenol; all preserve wood but their fumes are highly toxic to plants.)

The frame's front, sides and top are connected by means of ordinary door hinges; my friend can remove the hinge pins, take the pieces apart and store the whole thing flat against a wall of his garage. He brings it out for a couple of months each year in early spring and sets it up in a corner of the garden, with the top sloping toward the southern sun, to get his seedlings started. The sun generally keeps the temperature inside at an ideal 70° to 75° even on cool spring days; if the temperature goes much higher, which he can tell by glancing at a thermometer placed inside, he props the

USING A COLD FRAME

top open with a block of wood to allow the heat to dissipate. If a late spring frost is expected, he throws an old blanket or tarpaulin over the whole frame at night to conserve the heat built up in the soil and keep the temperature from dropping below 45°, the point at which most seedlings suffer damage.

Such a cold frame has other uses that not all home gardeners are aware of. In cool Northern climates it is sometimes difficult to get enough warm days to allow melons, eggplants and other long-season warm-weather crops to ripen before late summer, and one can rarely get early crops of cold-tender vegetables such as corn and beans. But seeds of these crops can be sown directly in the cold frame, and the plants can be allowed to grow right in the frame until they reach maturity. By the time the cornstalks or eggplants are tall enough to touch the cover, it can be removed entirely; melon vines can be allowed to wind around and around inside the frame and spill over the top. Of course if the frame is a portable one like my friend's, it can be removed entirely when the plants get too big for it. In any case, the cold frame makes it possible to have your vegetables on the table before frost gets them, and even if frost is no problem you will enjoy them weeks before plants in the open garden will mature. The frame comes in handy in the fall, too, for growing late crops of lettuce or radishes past their normal season, and

A COLLAPSIBLE COLD FRAME

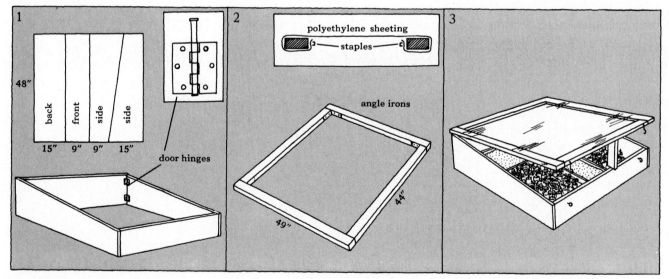

To make the base for a cold frame that can be taken apart for storage, cut a 4-foot square of ½-inch exterior-grade plywood as shown at upper left. Assemble the pieces with removable-pin hinges (upper right).

For the cover, fasten lengths of 1¼-by-3-inch lumber with angle irons. After painting, wrap clear polyethylene sheeting, 4 to 6 mils thick, over the top and secure it beneath with heavy staples (inset).

The finished cold frame, its cover hinged at the back with removable-pin hinges, can be propped open with a wood block for ventilation. Hooks and eyes in front will hold the cover securely in cold or windy weather.

even into winter for storing harvested crops of Brussels sprouts, celery and other vegetables that should be kept barely above freezing in order to last well and retain their flavor.

If you want to make still greater use of a cold frame, you can turn it into a hotbed by equipping it with a soil-heating cable so you can start seedlings directly in it during cold weather in early spring. Properly insulated cable, designed for the purpose and equipped with an automatic thermostat, is sold by many stores and mail-order houses. Remove the soil under the frame to a depth of 2 to 4 inches. Put down a 2-inch bed of sand, then loop the wire back and forth on top of the sand and cover it with a layer of ½-inch galvanized wire mesh, called hardware cloth, to protect the cable from damage by garden tools. On top of the mesh, add a 1- to 2-inch layer of sand. Pots or trays of seedlings can be placed directly on the sand, or if plants are to be grown directly in the soil of the hotbed, a 4- to 6-inch layer of soil can be substituted.

KEEPING DOWN WEEDS

Once your vegetable seedlings are up and growing in the garden, they will have to compete for moisture and nutrients with weeds, which will flourish in the rich, well-prepared soil every bit as well as the vegetables. The most common way of controlling weeds is to cultivate the soil, uprooting or cutting off the weeds while they are

SINGLE-STAKING TOMATOES

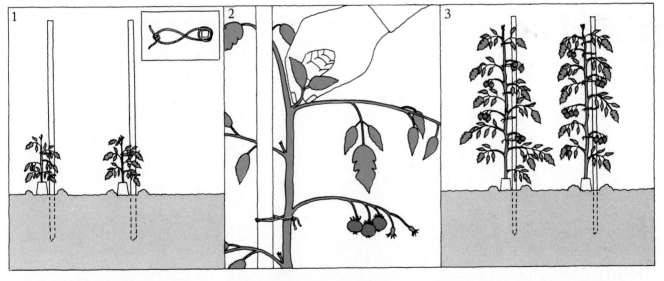

Tomatoes ripen a week or so ahead of schedule if trained to individual stakes. Plant seedlings 18 to 24 inches apart and drive a 6-foot stake 1½ feet deep beside each. Tie stems to stakes with soft cord (inset).

Limit each plant to a single main stem by removing leaf shoots that sprout in the joints between the leaf stems and main stem, but do not remove the flower- and fruit-bearing shoots between the leaf stems.

As the plants grow, keep the stems supported by tying them at 1-foot intervals, placing the ties just below the clusters of tomatoes. Single-staking not only speeds ripening but makes the crop easy to pick.

still small. Do it with a flat-bladed hoe, pulling the blade toward you so that it bites into only about the top ½ inch of soil. This depth is ample to decapitate the weeds: do not go any deeper or you will run the risk of cutting off the roots of the vegetables. Do not hoe too close to the vegetable rows; pull any weeds within the rows by hand. You can make weed pulling easier by moistening the soil an hour or so in advance so the weeds come out without disturbing the vegetables' roots. Hoeing for weeds can be combined with supplemental applications of fertilizer as required by different types of plants and specified in the encyclopedia section. Spread the prescribed amount in the strips between rows and it will be incorporated into the soil as you cull out the weeds.

USING A MULCH

To keep down weeds on a more or less permanent basis, and to allow the soil to retain moisture through dry spells, you can hoe or pull the weeds until the plants are about 4 to 6 inches high, then apply a mulch of straw or hay 3 to 6 inches deep around the plants and between rows. Grass clippings, applied not more than 2 inches deep because of their density, also make a good mulch. Such a layer will smother any further weed seedlings before they can get started, and by shading the soil, will make the need for watering less frequent. As it decays, the mulch will enrich the soil, but it will also settle and must be replenished. Some gardeners leave their mulch on year round for several years at a time, pulling it back only to plant. The soil under a permanent mulch, however, tends to compact from being walked on repeatedly, unless it is very light and sandy, and stays cold and wet longer in spring.

Whether or not you mulch, you may need to water your gar-

A PORTABLE VEGETABLE FENCE

A fence made of plastic garden netting not only supports vining plants such as cucumbers, peas and pole beans, but is easy to assemble in spring and take down for winter storage. For a fence 12 feet wide and 6 feet high, use three 8-foot metal posts—the kind that come studded with built-in hooks that can support the netting. Drive the posts into the ground at 6-foot intervals and loop the cords of the netting over the hooks on the posts. As the vegetables grow, the tendrils of the vines hold the plants tightly against the netting; for additional support in windy areas, tie the plants to the netting.

den during hot, dry weather. The best rule is to water, deeply and thoroughly, whenever plants start to show signs of wilting during the midday heat. To minimize the danger of fungus diseases, water in the morning or early afternoon so that wet foliage has a chance to dry out by nightfall.

Watering early in the day helps to ward off fungus diseases, but there are several other ways to protect vegetables from diseases and insects as well:

• Select disease-resistant varieties of plants; such varieties are noted in the encyclopedia and in most seed catalogues.

• Keep the plants growing vigorously in weed-free, well-enriched soil; large and healthy plants withstand attack better than small and weak ones.

• Pull up and relegate to the trash pile any diseased plant as soon as you spot it and before the disease can spread; do not put it on your compost pile. Do not handle a healthy plant after handling a diseased plant or you may spread the disease.

• To combat diseases that remain in the soil from year to year—and to prevent one type of vegetable from depleting the soil by extracting the same minerals again and again—try to rotate your crops, planting them in a different section of the garden each year. This is especially important for members of the cabbage family, as disease organisms to which they are prone as a group are carried over in the soil. An alternative is to fumigate the soil about a month before planting each year with an all-purpose soil disinfectant.

I do not believe in wholesale spraying, but in severe cases some application of chemicals offers the only alternative. Spray or dust with fungicides or insecticides targeted specifically at the problem you want to control, as shown in the charts on pages 152-154, and follow all the precautions printed on the labels.

As the pest chart indicates, any spraying of vegetables should be stopped from several days to several weeks before harvesting, depending on the vegetable and type of chemical, to prevent any harmful residues from finding their way to the dinner table; the recommended intervals are usually specified on the labels. Then your only remaining task is a pleasant one: picking the harvest and preparing it the way you like. I love most fresh vegetables, eaten raw or prepared in many ways to bring out the best in them. I guess if I had to name one favorite it would be fresh sweet corn. Of a summer evening, Margaret gets the water boiling and I go out into the garden to pick a few ears; if I don't move quickly I soon hear about it, for I have often pointed out how rapidly corn can lose its sweet, milky taste. We pop the ears into the water for three and a half minutes—no more, no less—then serve it immediately, with mounds of fresh butter, pepper and salt. Need I say more?

A PLASTIC ROOT CELLAR

The old "root cellar," where great-grandmother stored carrots and other root vegetables over the winter, has almost disappeared from American homes. But a plastics-age version is easily installed where winter temperatures average 40° or lower; it is simply a garbage can buried in the ground. Sink a 32-gallon can of heavy-duty plastic with the rim about 3 inches above soil level so that rain and melting snow cannot leak inside. To prevent the vegetables from shriveling, line the bottom of the can with 2 inches of damp sand and pack more damp sand around and between the layers. The top is insulated against changing air temperatures by a readily movable covering made up of a 1- to 2-foot-high mound of straw or leaves kept dry by a piece of plastic film; the film is held in place by rocks.

An expert's vegetable garden

Alexis Bervy, who retired to Florida a few years ago, raises vegetables as a hobby, but he practices his avocation with the skill of an expert as well as the love of an amateur. The results (right) would be the envy of any neighborhood: succulent tomatoes, firm green sweet peppers, tasty onions, crisp snap beans and carrots, plump eggplants that frequently weigh up to 2 pounds each.

Bervy has a jump on the average home gardener: before moving to suburban Miami he had a small truck garden on his dairy farm in upstate New York, and he worked out his techniques over a period of years. In Florida he is blessed with a longer growing season, but his basic methods remain unchanged, and they can be adapted with profit by home gardeners anywhere. The bounty and quality of his harvests are not so much the products of hard labor—he spends only a few hours a week on his 22-by-42-foot garden—as they are of meticulous preparation and care. He raises all his vegetables from seeds rather than from nursery-started plants, a method that not only is cheaper but gives him a chance to experiment with a wider range of varieties, including new ones as they appear in the seed catalogues. A month before planting he digs the soil about 8 inches deep, incorporating a soil-disinfecting fumigant to control diseases and weed seeds. A week before planting he rakes in a low-nitrogen high-phosphorous chemical fertilizer and repeats the fertilizing every two weeks during the growing season. To be sure his vegetables receive ample moisture, Bervy installed a simple sprinkler system (overleaf), with which he gives his garden an inch or more of water a week during hot weather. He waters early enough in the day so that the foliage is dry before sundown, thus avoiding conditions that can lead to fungus diseases.

Bervy's systematic approach pays off dramatically. He planned the garden to be compact enough to handle easily on his own, yet big enough to feed a family of four. In fact, under his skillful hand, the plot produces so abundantly he estimates that it could provide year-round eating for 10 people or more.

The day's picking done, a smiling Alexis Bervy kneels behind a bushel basket brimming with 11 different vegetables.

A well-placed, well-watered plot

Alexis Bervy knows that vegetables need as much light as they can get for proper growth. For this reason he located his garden well away from the fruit trees that dot his property so it can receive 8 to 10 hours of full sun a day. To avoid the constant need for reeling and unreeling a hose, he built a sprinkler system out of galvanized pipe sunk in trenches well belowground so it does not interfere with cultivating. He placed the 4-foot-high sprinkler poles so that their spray patterns overlap, reaching the vegetables in every row.

The Bervy garden is only a few steps from the house, so baskets of produce do not have to be carried far. At either end, posts and wire support tomato and cucumber plants.

A plan for planting, picking and preserving

Like other knowledgeable vegetable growers, Bervy keeps an annual record of his garden, noting the varieties of vegetables, the number of plantings and harvests, and the total production figures in terms of fresh vegetables harvested and the amounts that were preserved or frozen. The only permanent fixtures in the garden are the wire supports at either end, on which he grows vining plants to save space and keep the fruits off the ground. The rest of the garden Bervy changes around at will. The diagram below and the notes at right show a typical year's plan. Because of the long growing season, Bervy is able to make up to five successive plantings of crops like snap beans and cabbage.

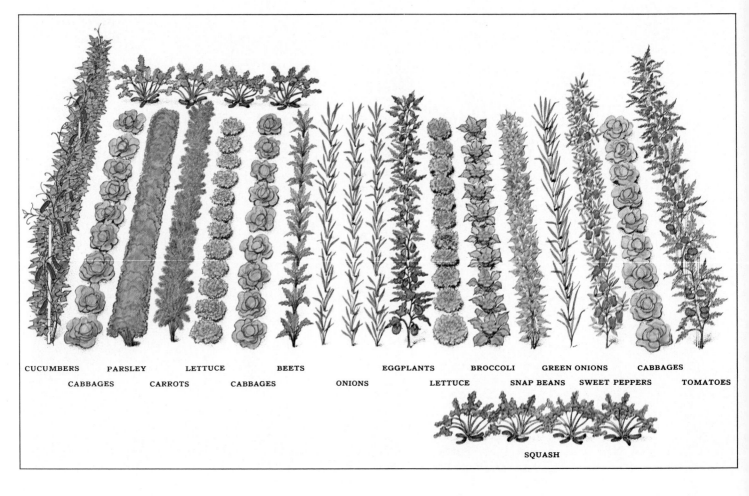

CUCUMBERS   PARSLEY   LETTUCE   BEETS   EGGPLANTS   BROCCOLI   GREEN ONIONS   CABBAGES

CABBAGES   CARROTS   CABBAGES   ONIONS   LETTUCE   SNAP BEANS   SWEET PEPPERS   TOMATOES

SQUASH

CUCUMBERS

Variety: Poinsett. One row of 40 plants replanted twice during a six-month period. Cucumbers mature in three months and are picked two or three times a week continually over a six-month harvesting season. Yield per row per planting: 120 pounds. Preserved: 12 quarts of pickles.

CABBAGES

Variety: Early Jersey Wakefield. Three rows, 22 plants each. Seeds continually planted to replace harvested heads, which mature in three months and are picked weekly over a six-month season. Yield per row per planting: 22 heads (3 to 4 pounds each). Preserved: 8 quarts of sauerkraut.

PARSLEY

Variety: Dwarf Green Curled. One row, 22 plants. Plants mature in about three months, although some sprigs can be picked after two months. One planting produces for about six months, providing ample parsley for steady use as a garnish and in making pickles.

CARROTS

Variety: Chantenay. One row planted, replaced by another later as needed. Seedlings thinned to leave about 250 plants 1 inch apart. Carrots mature in about three months; two plantings suffice for the six-month season. Yield: 45 one-pound bunches per planting.

LETTUCE

Variety: Great Lakes. Two rows, three successive plantings of 22 plants per row, 10 inches apart. Early heads ready in about three months; mature heads picked over four months. Yield: 22 heads per row per planting.

BEETS

Variety: Detroit Dark Red. Two plantings provide about 120 plants per row, 2 inches apart. Beets are ready in two months and are picked weekly during a four-month harvesting season. Yield per row per planting: about 27 one-and-one-half-pound bunches. Preserved: 10 quarts.

ONIONS

Varieties: Southport Yellow Globe (two rows), Sweet Spanish (one row). One planting of each, grown from seed and thinned to about 100 plants per row. Both varieties are ready in about three months and are harvested over a six- to seven-month period. Yield: 30 pounds per row.

EGGPLANTS

Variety: Black Beauty. One row of six plants, set 18 inches apart. Begin to yield after two and a half months; harvesting continues for more than six months. Yield: 15 to 25 eggplants, each 1 to 2 pounds. Preserved: 12 quarts of eggplant relish.

BROCCOLI

Variety: Calabrese. One row of 10 plants, 18 inches apart. Matures in about three months; heads are picked two or three times a week over a seven-month season. Yield: 10 two-pound bunches. Frozen: 10 one-quart packages.

SNAP BEANS

Variety: Harvester. Seeds are sown in groups every two weeks for one row of 120 plants 2 inches apart. Beans mature in two months; when plants stop bearing they are replaced by new seeds. Total harvest period: seven months. Yield: about 25 pounds. Frozen: 20 one-quart packages.

GREEN ONIONS

One row of about 200 onion sets, or small bulbs, is planted and picked in six weeks for eating as green, or bunching, onions. Bulbs not picked are matured for use as ordinary large onions. Yield: 20 pounds (64 five-ounce bunches).

SWEET PEPPERS

Variety: California Wonder. One row of 12 plants, spaced about 18 inches apart. Peppers begin to mature in three months; plants produce for six months. Yield: about 450 peppers. Preserved: 10 pints of sweet pickled peppers.

TOMATOES

Variety: Manalucie. One row of 15 plants, replanted twice over a six-month period. The first tomatoes mature in three months; each planting will produce for two months. Ten pounds picked twice a week. Yield: about 75 pounds. Preserved: 10 quarts of whole tomatoes, 10 pints of chili sauce.

SQUASH

Variety: Hybrid zucchini. Two clumps, or hills, of three or four plants each sown at either side of the garden. They begin to mature in about two months and are picked over a seven-month period. Yield: about 75 squashes per hill.

The Poinsett variety of cucumber was chosen by Bervy because it grows especially well in the South, resisting disease and yielding abundantly—and because he liked its crisp flavor. Harvested when only 4 to 5 inches long, it can also be made into pickles.

Early Jersey Wakefield has long been a favorite cabbage of the Bervys, who enjoy it in coleslaw, freshly shredded and dressed with a spicy mayonnaise sauce, and in Mrs. Bervy's special sauerkraut, which is made of chopped cabbage layered with apples and salt.

Calabrese broccoli is Bervy's choice for mild taste and high yield. When the large central head of each plant is cut, many smaller heads keep forming on the side branches; one planting of Calabrese lasts the Bervy family for about six or seven months.

Among eggplant varieties, Black Beauty was chosen because it produces large and tender fruit; the fruit also stay on the vine when mature instead of dropping to the ground, making picking easier. Most of the crop goes into Mrs. Bervy's eggplant relish (overleaf).

Detroit Dark Red is Bervy's choice among beets because it is disease resistant and keeps its fine-grained texture even when fully grown. He likes it, too, because it holds its sweet flavor whether it is canned, frozen or freshly cooked along with the tasty beet greens.

The variety of sweet pepper known as California Wonder was selected by Bervy because each plant yields up to two dozen peppers during the growing season. Mrs. Bervy considers its blocky shape and thick walls ideal for stuffing, and its firm flesh good for pickling.

Favorite varieties with key qualities

Alexis Bervy makes a methodical selection from among the many vegetable varieties, new and old, when he plans his plot each year. First he narrows his choices to varieties that will grow well in this area, checking ones he is not sure of with the county agent. After growing them for a year he rates these varieties by two key qualities: flavor and productivity. He likes the Manalucie tomato, shown below, on both counts because it not only bears a heavy crop but also has a high proportion of tasty edible flesh in relation to juice and seeds.

Mrs. Bervy cuts sweet peppers for her eggplant relish into strips before feeding them through the grinder. The tomatoes will follow, then trimmed and cut-up scallions.

Ingredients for the relish, all prize vegetables from the Bervy garden, are assembled on the kitchen table. At right are jars of the product, ready for storing or giving to friends.

Mrs. Bervy's eggplant relish

Dunia Bervy is somewhat of a local celebrity for her eggplant relish, made from an old family recipe. To make 10 quarts, she bakes four eggplants for an hour at 400° and grinds up 10 tomatoes, seven sweet peppers and 5 cups of chopped scallions. She peels and chops the eggplants and mixes the ingredients in a large bowl with 1 cup of salad oil, ¾ cup of vinegar, 3 teaspoons of salt and 1½ teaspoons of pepper. Fresh relish is allowed to mellow for a day or two before use; preserved relish is sealed and boiled in jars for 45 minutes.

How to raise fruits, nuts and berries 3

I grew up on a fruit-producing farm in northern Massachusetts and have been a devoted fruit grower ever since. Long before my father bought the farm, I began to learn about the ways of fruits from the previous owner of the place, old Mr. James. I can see him still, trudging down the road each spring with the tools he used for grafting fruit trees, for his services were in constant demand by other farmers in the area. While I was still a boy he taught me how to tuck a twig from one variety of apple tree into a cut in another and transform an inferior tree into one that bore mouth-watering fruit. Mr. James practiced what he preached: his own orchard had more than a hundred different varieties of apples alone, as well as peaches, pears, cherries and walnut trees. And besides the trees, there were strawberries, raspberries, blackberries, currants, gooseberries and grapes. There was no such thing as a cultivated variety of blueberry in those days; we picked ours wild in the pasture, sold some and ate basketfuls ourselves.

A lot of things have changed since those days. There have been an incredible number of improvements in the varieties of fruits available for growing, in fruit size, productivity, taste, and resistance to disease and cold. You used to have to climb tall ladders to prune and harvest apple trees; today you can perform these chores with both feet on the ground because modern dwarf types grow only 6 to 8 feet tall. In the old days it was difficult to grow peaches north of New Jersey; now there are varieties that tolerate temperatures of 25° below zero and produce fruit as far north as New Hampshire. At least in part because of these developments, many people are discovering, or rediscovering, the old-fashioned pleasures of growing fruits, nuts and berries: the fun of picking your own right off the bush or tree; the knowledge that fresh fruits provide all sorts of vitamins, notably A and C, that are partly lost in stored or processed food; and the ability to satisfy a sweet tooth when you want to without putting on a lot of weight.

Of all the fruits available to the home gardener, the various

A basket of ripe strawberries, America's favorite berry fruit, nestles among flowering and fruiting plants in a Maryland experimental garden. The varieties pictured here are hybrids being tested.

types of berries are the most widely grown, and not only for their taste. Among the other reasons: the best types do not ship well and are often unavailable in stores; berry plants produce a great deal of fruit for the space they occupy; and berries are easy to grow and relatively free of pests and diseases. Of all the berries, by far the biggest favorite is the strawberry, which in one variety or another will flourish in every part of North America. Strawberries are so well loved in the United States, in fact, that many people think of them as the native all-American fruit. Actually the modern strawberry originated in France in 1766 as a cross between the North American wild strawberry, *Fragaria virginiana,* and the wild strawberry of Chile, *Fragaria chiloensis.* The first hybrid, larger and tastier than either parent, was subsequently bred and rebred to produce the still-sweeter, plumper berries we enjoy today. The Latin genus name, *Fragaria,* refers to the delightful fragrance of the fruit, the common name to the fact that growers keep the fruit-heavy stems off the ground with a mulch of straw.

Almost as popular—and for much the same reasons: good taste and perishability—are raspberries, blueberries and blackberries of various types. About the only problem with the berry fruits is the fact that birds love to eat them just as much as people do, but it is a simple matter to throw plastic netting, made for the purpose, over the plants as the berries start to turn color; the birds may get a few through the netting but you will be able to save most of the crop for yourself.

I enjoy all these small fruits, but I have a special spot in my heart for that most celebrated of vine fruits, the grape. One reason, I suppose, is that I live not far from where the famous Concord grape was born and have along the back fence two Concord vines, direct descendants of the original vine. That vine, bred from wild grapes in 1843, was so superior to varieties in use at the time that now, more than a century later, it and the hybrids bred from it are still standards by which other table grapes are measured. Home gardeners all over the country still grow Concords for their sweet blue-black fruit, which make fine eating as well as superb grape juice, jelly and jam. Some gardeners hesitate to grow Concord and other kinds of grapes because they think grapes require a lot of work; actually they are quite easy to grow, if you know how and when to prune and tie them in order to channel their normally rampant growth into the production of large, healthy grapes rather than luxuriant stems and leaves *(drawings, page 52).*

THE TREE FRUITS

Rewarding as grapes, bush fruits and strawberries are, the crowning glories of any garden, mine included, are its fruit and nut trees. The most popular nut trees—pecans, walnuts, Chinese chestnuts

—not only produce delicious nuts in fall, but are generally long-lived and handsome assets in the home landscape and are especially prized as shade trees all summer long. Chinese chestnuts and black walnuts in particular are fast growing, bearing nuts as early as two years after planting and casting useful shade within five or six years, and they are relatively free of pests and diseases.

Fruit trees require a little more work than nut trees and the smaller berry plants in the way of pruning, picking and pest control, but they are worth every bit of the effort involved. This effort has been diminished in recent years by the breeding of disease-resistant varieties, but even more important has been the development of the dwarf fruit tree. Naturally stunted trees, called genetic dwarfs, had been grown as novelties on and off for some 300 years. In 1912, horticulturists at the East Malling Research Station in Kent, England, began to collect as many as they could find for experimentation. They discovered that a standard apple variety grafted to the roots or lower trunk of a dwarf would itself become dwarfed, growing to a height of 8 to 15 feet instead of the normal 25 feet or more. Because some dwarfing rootstocks develop limited root systems that are too small to anchor the trees firmly in the ground and might not survive cold winters, several American growers now use stem sections taken from dwarfing stock and graft

(continued on page 49)

PLANTING STRAWBERRIES

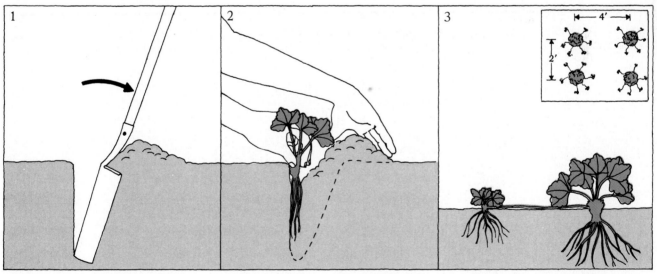

To accommodate the long fan-shaped roots of a strawberry plant, make a V-shaped planting hole by sinking a flat-bladed spade vertically into the soil, about 8 inches deep, and pulling the handle toward you.

After shaking the plant to untangle and spread its roots, hold it against the side of the hole so that half the stem base, or crown, is above soil level. Push soil against the roots, then firm it with your foot.

Each plant sends out runners that sprout new plants. To arrange your bed in spaced-matted rows (inset), in which plants bear abundantly and require little care, cut off all but six evenly spaced, vigorous runners.

A new interest in old-time apples

If you remember the apples of your childhood as being more flavorful and fragrant than today's store-bought types, you are not just being nostalgic. Hundreds of varieties of mouth-watering apples that once flourished in backyard gardens and orchards have nearly disappeared, having proved unsuitable for large-scale marketing. But due to the efforts of specialists like S. Lathrop Davenport (*below*), an increasing number of home gardeners are raising their own Sweet Winesaps, Rhode Island Greenings and other old-fashioned varieties from stock grafted onto ordinary apple trees.

In a Massachusetts apple orchard, S. Lathrop Davenport displays some of the 60 varieties of old apples he has helped to revive by searching them out all over New England. In his right hand is a rough-skinned but sweet-tasting American Beauty apple.

The crisp juicy Sweet Winesap, which originated in Pennsylvania around 1850, was prized for desserts in the days when sugar was in short supply.

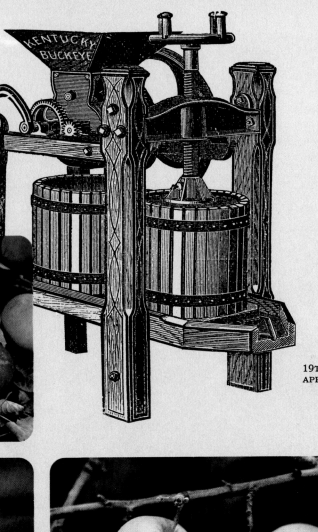

19TH CENTURY
APPLE CIDER PRESS

The yellowish, distinctively striped Tolman Sweet was developed in Dorchester, Massachusetts, in the late 1800s. Once rated the best apple for wintertime eating, it is also excellent sliced and baked in pies and cakes.

The curiously colored Blue Pearmain is bright red with a bluish cast and yellow markings. Like many of the old apple varieties, it is aromatic and sweet despite its somewhat hard, rough skin and unappetizing appearance.

47

The deep red, fragrant Ben Davis, long popular because it bears abundant amounts of fruit that keep well throughout the winter, was brought up from the South to New England in the early 1800s.

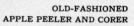

OLD-FASHIONED APPLE PEELER AND CORER

The large Rhode Island Greening was developed in the early 1700s near Newport. Its tender tart flesh was often preserved by drying as well as used fresh in cooking.

these sections between a sturdy, standard-sized rootstock and the desired variety of fruit to achieve the same effect. To get even better results, one grower I know literally builds some of his dwarf apple trees out of four rather than three separate components. He uses a sturdy French crab apple for the initial rootstock, grafting onto it below ground level a cold-resistant apple variety, which puts out some of its own roots as well. While this two-part tree is growing in the nursery he assembles a second combination, splicing the desired fruit-bearing variety on top of a stem section of a dwarfing variety. Finally, he grafts the two assemblies together into a single apple tree that is shipped to the customer after it has established itself for two years in the nursery.

Because of the skilled labor and growingtime involved, dwarfs cost a bit more than standard fruit trees of the same type, but the results they produce are truly remarkable. The trees bear fruits one to three years after planting, compared with 5 to 10 years in the case of some standard apple varieties. A dwarf produces, on the average, 1 to 3 bushels (50 to 150 pounds) of fruit per season, ample for most home gardeners, who generally do not know what to do with the 15 to 20 bushels of fruit produced by a standard tree; moreover, the fruit are every bit as large as that of a standard tree, and often they are larger. Because dwarfs grow only 6 to 8 feet high —15 feet in the case of semidwarfs—they are easy to take care of with ordinary garden spray equipment and can be pruned and picked from the ground or a short stepladder. They also require much less garden space: in the amount of ground area taken up by a mature standard apple tree you can plant a half-dozen dwarfs. Peach, plum and cherry trees are also available in dwarf form, but because standard-sized trees of these fruits do not grow as large as apple trees the advantages are not as great.

Regardless of their size, fruit and nut trees, like berry fruits and vegetables, should be chosen with a careful eye to varieties that will grow well in a given region. Apples, and even some peaches and plums, cannot be grown in the Deep South because they do not get long enough periods of cool temperatures, which they need for a rest period each year if they are to grow properly and produce fruits. Conversely, citrus trees cannot be grown in the North because the winter cold kills them. Some varieties of pecan trees can be grown quite far north, but will not bear well-developed nuts because temperatures do not stay high enough over a long enough growing season to allow the kernels to mature to full size. The encyclopedia section and the zone map indicate which varieties flourish in your region.

Even when a fruit or nut tree is planted in the proper zone, a late spring frost may kill its flower buds; such a loss is particularly

OLDTIME APPLES TO GROW

Gardeners interested in growing some of the delicious apples of yesteryear (pages 46-48) can obtain young semidwarf or dwarf trees of many varieties, or stock to graft onto their own existing trees. These are available through preservation orchards such as the one in Old Sturbridge Village, Massachusetts, or from specialists who can be reached through local nurseries, horticultural societies or county agricultural extension services. Among the most popular old-fashioned apples, listed according to their maturing season, are the following:

SUMMER
Benoni, Early Harvest, High Top Sweet, Yellow Transparent.

LATE SUMMER-EARLY FALL
Cox Orange, Alexander, Porter.

LATE FALL
Ben Davis, Black Gilliflower, Blue Pearmain, Esopus Spitzenburg, Fameuse, Lady, Maiden's Blush, Pound Sweet, Rhode Island Greening, Roxbury Russet, Sweet Winesap, Tolman Sweet, Twenty Ounce, Wagener, Yellow Newtown.

likely with early-flowering trees—peaches, apricots and hazelnuts, for instance. For this reason the best site for most fruit trees is a north-facing slope, where the sun will not stimulate the buds to grow as soon as it would on a site facing south and where the slope will allow the heavy cold air of frosts to flow quickly away down-hill instead of accumulating to nip the buds.

Fruit trees, like berry plants, need full sun and well-drained soil. But unlike most berries, many fruit and nut trees—including most apples, pears, plums, cherries, walnuts and pecans—cannot be fertilized by their own pollen; they will remain barren and fruit-less unless there is a tree of the same kind but another variety near-by to ensure cross-pollination. If you want to plant an apple tree such as a McIntosh, which flowers in early spring and bears fruit in late summer, make sure there are some other varieties of apples around so that the bees visiting their flowers will deposit the pollen in the flowers of your tree. If one of your neighbors happens to have an apple tree, you have no worries; otherwise, you can always plant two different varieties yourself.

PLANTING AND PRUNING

The best fruit trees to plant are ones that are two years old and 3 to 5 feet tall; not only do they survive transplanting better and take hold faster than older trees, but they will actually outgrow

PRUNING RED RASPBERRIES

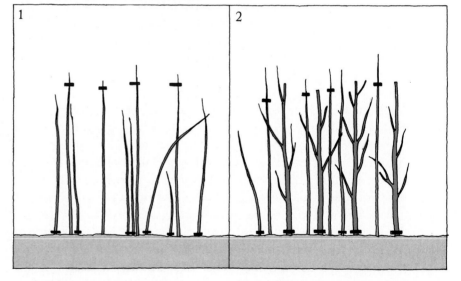

In early spring of their second year, thin red raspberry canes (bottom lines), leaving two or three per foot of row. Prune to 3½ feet (top lines) to force the growth of side branches, which fruit that summer.

Each spring cut away the branched canes that bore fruit the preceding year and thin the new canes, removing imperfect, crowded ones and top-pruning the remaining two or three per foot of row.

them and begin bearing sooner. Moreover, a young tree can be trained properly from the moment it is planted to look shapely and bear heavy crops without having its branches break under the weight of the fruit or the lashing winds of summer storms. Most fruit trees, with the exception of citrus, are sold and planted while they are leafless and dormant, when they are easily dug, shipped and handled with their roots bare of soil. In Northern areas (Zones 3-5) most fruit trees should be planted in early spring, as soon as the soil can be worked, to give them a summer in which to become established before fall and winter set in. In Zones 6-10, fruit trees can be planted in spring, but because summers are likely to be long and hot, it is preferable to plant them in fall or winter; at that time the soil, either unfrozen in Zones 8-10 or lightly crusted with frost in Zones 6 and 7, is still warm enough to promote the development of strong roots before the trees start top growth in spring.

Most knowledgeable gardeners buy their fruit and nut trees from a reputable mail-order nursery specializing in fruit plants (the majority of local nurseries do not stock them). The young trees arrive with their roots wrapped in damp packing material and plastic. The roots must not be allowed to dry out or the tree will die; leave the tree in the packing in a cool, shaded place, then the day before you intend to plant set the roots in a pail of water to soak over-

TRAINING TRAILING BLACKBERRIES

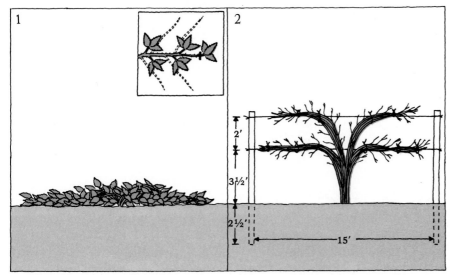

Newly planted trailing blackberries, after being cut back to 6 inches, send canes growing along the ground. When they reach 8 feet snip off their tips (solid line in inset) to force the growth of laterals (dotted lines).

Early the next spring, cut away all but 16 of the leafless canes. Tie them to a two-wire fence in bunches of four canes each and cut the laterals back to 1 foot. After they have borne fruit, cut the canes to the ground.

night. Make a generous-sized hole, at least half again as large as the spread of the tree's roots *(drawings, page 54)*. Pile topsoil and subsoil separately, and use the richer topsoil in the bottom of the hole to nourish the roots. To loosen the soil and give it good moisture retention and drainage, mix in 1 part peat moss to every 2 parts of topsoil and subsoil. Most soils are of a pH suitable for growing fruit trees; to get optimum growing conditions, extremely acid or alkaline soils can be adjusted by adding limestone or sulfur *(Chapter 1)*, although it must be dug deeply into the ground in the area around the tree where the roots will spread and it may have to be re-applied every few years if the soil reverts to its original level of acidity or alkalinity. Do not add any fertilizer when planting; there are enough nutrients in most soils to stimulate normal growth, and the

PRUNING AND TRAINING AMERICAN GRAPES

1. *After the vine has been planted and cut back to two buds (inset) the buds will sprout laterals, or side branches; cut off the weaker one (line) and tie the other to a 5-foot stake sunk 1 foot into the ground. This first-year pruning is done in all grape-pruning systems.*

2. *The following spring, train American grapes by the four-cane Kniffen system, tying the vine to the top wire of a two-wire fence and cutting off the tip 3 inches above the wire. For training European grapes, see page 53.*

3. *The third spring, tie four laterals to the wires (a) and prune them to six buds each; these will produce the summer's crop. For next year's fruit, prune four nearby canes to make "renewal spurs" with two buds each (b, and inset). Cut other canes away.*

4. *The fourth spring, cut off the tied canes that have borne fruit (a, and inset). Tie up one new cane from each renewal spur and cut it back to six buds (c). Cut back the other new canes to two buds each; they form renewal spurs for next year's fruit (d, and inset).*

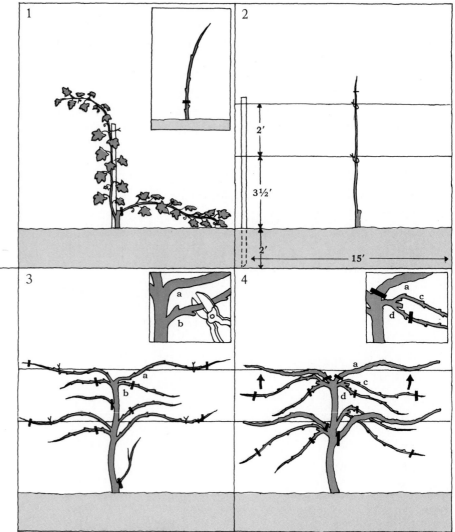

newly developing roots are extremely sensitive to overfeeding and burning by chemicals. Fertilizer, if any is necessary, can be given after the roots are established, as described in the encyclopedia. If you are planting a dwarf tree, make sure to keep the crooked, swollen graft at the base of the trunk aboveground, otherwise roots may form on the standard-sized variety above the graft and bypass the dwarfing rootstock, allowing the tree to grow to the full height of the standard type.

Probably the most difficult thing for the average gardener to grasp is the next step: removing about half the top growth. Having paid for a 5-foot tree, you might logically think you are entitled to a 5-foot tree in your garden right away. But when the tree was dug at the nursery, about half of its root system was inevitably lost, and an equal amount of top growth must be removed to bring top and bottom into balance again; otherwise the tree will grow feebly if at all, and branches may die back. Of course the need for pruning does not mean you should whack off half the top; such butchery would not only make for a very unsightly tree but might be as fatal as no pruning. The object of initial pruning is to remove branches that total about half the top growth, starting with weak and poorly placed ones. The result should be a tree that has three to five strong, well-distributed "scaffold" branches, as shown in the drawings, from which new growth will soon come.

After planting and pruning, water the young tree thoroughly and give it about a pailful of water every week during the first summer except when an inch or more of rain has fallen. To retain moisture in the soil and keep down competing weeds, spread a mulch of ground bark or wood chips 2 to 3 inches deep around the trunk, within a shallow saucer of earth made to hold water.

Many gardeners make the mistake of applying too much fertilizer to young trees, after planting as well as before, in the belief that they will get bigger, better fruits at an earlier date. While I believe in feeding shade and ornamental trees well, I feed my fruit trees rather sparingly, and I scatter the fertilizer on the ground rather than placing it in deep holes as I do with other trees. The main reason is that fruit trees, if fed too well, grow branches and leaves at the expense of bearing good crops of fruits, and they may actually start bearing at a later age than those allowed to develop at a normal pace. A more effective method of inducing a tree to bear fruit sooner in its life is the time-honored one of bending its branches over so that they grow into a horizontal or near-horizontal position. Bending constricts the sap-carrying layer of the branch enough so that more of the nourishing carbohydrates manufactured in the leaves stay in the branch—stimulating the formation of extra blossom buds and thus more fruit. One way of training branches

PRUNING EUROPEAN GRAPES

The easiest way of training European grapes is the spur-pruning method, which produces vines that look like this at the beginning of their third season. The first season the vine is cut back to two buds after planting; the same summer remove the weaker of the two canes that sprout. The second spring tie the vine to a 2- to 3-foot stake and cut the vine off near the top of the stake, forcing side canes to grow. The third spring remove all but three to six of the topmost canes, and cut back those remaining canes to two buds each (lines); the canes that grow from these spurs will bear fruit. Each spring thereafter, prune away all sucker branches arising from the lower trunk, but retain an increasing number of spurs at the top each year.

PLANTING AND PRUNING A FRUIT TREE

1. *To plant a fruit tree, dig a hole larger than the spread-out roots and fill it about one third full with 2 parts topsoil and 1 part peat moss; the crooked graft on its trunk should be 2 inches aboveground (or 2 inches below if the tree is standard-sized rather than a dwarf). Fill the hole two thirds full with soil mix and firm with your feet (inset).*

2. *Water thoroughly, then fill in soil mix to the top; build a 2-inch-deep earthen saucer (inset) around the tree and water again.*

3. *Prune off all but three to five strong branches; they should be 6 to 12 inches apart, spread in different directions and form a 45° angle with the trunk. On most trees, prune these branches to 8- to 10-inch stubs, cutting ¼ inch above an outward-facing bud (inset). On peach and apricot trees, however, prune the branches back to 2 to 4 inches.*

4. *Protect the bark with tree-wrap tape, and tie the tree—using wire threaded through garden hose (inset)—to a 6-foot post driven 2 feet into the soil.*

5. *To shield the trunk against rodents, encircle it with a 2-foot-high collar of the ½-inch wire mesh called hardware cloth, sunk 1 to 2 inches into the soil. Fill the watering saucer around the tree with 2 to 3 inches of mulch.*

6. *In the spring following two seasons of growth, prune the tree again to develop an open airy structure that lets light and air reach all of the fruit. Remove crowded, low-growing and new competing branches, making cuts above outward-growing buds or shoots to avoid crossing branches.*

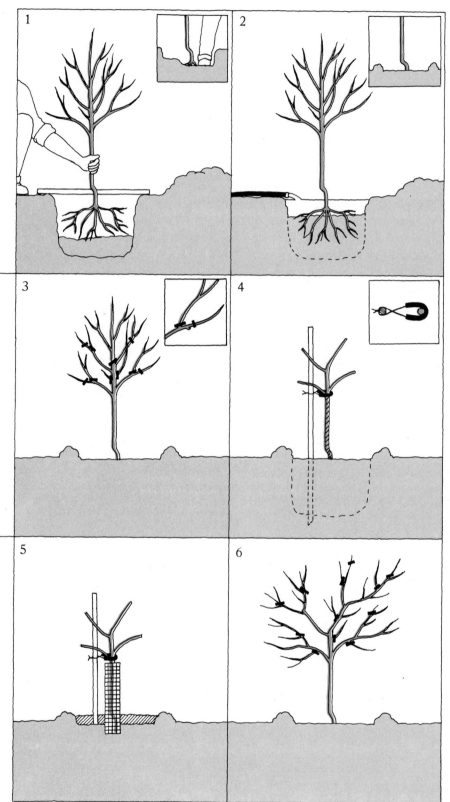

to grow horizontally is to bend the branches down as far as they will comfortably go without cracking, and to tie them in this position with heavy twine attached to stakes driven into the ground. The ultimate expression of this technique is the ancient art of espaliering, which trains a tree to grow with its branches spreading horizontally against a fence or wall *(drawing, right)*.

Regardless of how a fruit tree is fed or trained, however, you will not get a satisfactory crop from it unless you pay some attention to pruning, fruit thinning and pest control. After the tree's initial pruning at the time of planting, further pruning should be restricted to light shaping as shown in the last drawing on page 54: any heavier pruning simply delays fruit production. When the tree does reach bearing age, you will have to face one of the horticultural facts of life: to get good fruit, unblemished by insects and diseases, you will probably have to spray. I once complimented a friend of mine who has a commercial peach orchard, remarking on the magnificent quality of his fruit. "It *should* be good," he replied, "I sprayed my trees 14 times this year!" Not all home gardeners have the time or inclination to keep up such a rigorous schedule of pest control, and actually not all the sprays are necessary if you are willing to accept what the commercial grower cannot afford to: some less-than-perfect fruit. A full schedule of sprays, with the essential ones noted, is given in the chart on pages 150 and 151.

With proper pruning, feeding and spraying, most fruit trees will bear abundantly—sometimes too abundantly. A heavy crop of fruit lessens the size of individual fruit and frequently exhausts trees, causing them to bear crops every other year, with only a scattering of fruit in off years. To prevent this tendency it is advisable to thin the fruit, cutting down the total number but distributing them more evenly on the branches to allow more leaves to nourish each; the result is larger, more brightly colored fruit, often with better flavor. Thinning is best done when fruit are about one third grown, so that the remaining ones are 2 to 8 inches apart depending on the type *(encyclopedia)*.

When the fruits ripen, there are only a couple of final things to keep in mind. To make sure you get unbruised fruits that will taste their best and keep for the maximum amount of time, treat them as though you were handling eggs; even though a peach, plum or apple may feel firm, its skin is fragile and if broken is quickly susceptible to rot. Pick one fruit at a time, not several; and leave the stems attached so you do not open the tops to infection. Set the fruits as you pick them in a basket padded with a folded cloth, and store them in the refrigerator or the coolest place in the house. With just a little care, you can enjoy them at their peak of flavor for weeks to come.

AN ESPALIERED FRUIT TREE

Fruit trees such as the dwarf pear shown here can be trained by the espalier method to form attractive and productive patterns flat against a sunny fence or wall. To make the pattern illustrated, a one-year-old unbranched tree is planted and cut back to about 15 inches, just above two buds growing in opposite directions. As branches sprout from the buds, they are bent down and tied to wires or trellis slats. When the trunk tip grows 12 to 15 inches higher, two more buds are chosen and the process is repeated. Other trunk shoots are removed as they appear and upright branch shoots are pruned just above their third leaves (inset, upper lines); stubby fruit spurs are left alone, but the growing fruit is thinned (lower lines) as specified in the encyclopedia.

The grete herball

which geueth parfyt knowlege and vnder
standyng of all maner of herbes & there gracyous vertues whiche god hath
ordeyned for our prosperous welfare and helth/for they hele & cure all maner
of dyseases and sekenesses that fall or mysfortune to all maner of creatoures
of god created/practysed by many expert and wyse maysters/as Auicenna &
other.&c. Also it geueth full parfyte vnderstandynge of the booke lately pren
tyd by me(Peter treueris)named the noble experiens of the vertuous hand
warke of surgery.

The practical and pleasing herbs 4

Thirty-two jars of dried herbs line the shelves of the open cabinet above the counter where my wife Margaret cooks. Across the kitchen, pots of chives, rosemary and parsley have displaced the begonias and geraniums that occupied our sunny window ledge for years, and though the herbs grow luxuriantly they never seem to get bigger because so many of their leaves are snipped off to add zest to everyday meals. Margaret, like so many modern cooks, has discovered what a few herbs can do and she puts a snippet of one or another into everything from scrambled eggs to stuffed potatoes.

While Margaret prizes herbs for their contribution to cooking, I welcome them in the garden, where the perfumes from their blossoms mingle with the fragrance of their leaves. Most of them grow like weeds, with a minimum of effort on my part; and though many of these herbs require a sunny corner, a few such as mint and chervil will flourish in the shade.

The term herb, of course, is applied to hundreds of plants, including purely aromatic and decorative ones (like those shown in the picture essay that begins on page 64). But I find that the culinary kinds offer the greatest rewards. All 23 herbs listed in the encyclopedia section are easy for weekend gardeners to grow in virtually any part of the country, and all of them can be used to season a wide variety of foods *(page 61)*.

Since most culinary herbs are natives of the dry regions of the Mediterranean, the secret of growing them successfully in your garden is to give the plants good drainage. Do not water unless the plants wilt noticeably during the hot part of the day. Neither is fertilizer necessary very often; far from needing rich loam, herbs usually grow most vigorously and develop their best flavor in thin infertile soil. The only care most of them need is weeding. Diseases rarely trouble them, and their scent seems to repel insect pests.

Methods of growing herbs differ according to their type: that is, whether they are annuals, biennials, perennials, perennial bulbs or shrubs. Some of the commonest herbs are annuals, which grow

Herbals, books that prescribe herbs for all sorts of medicinal uses, were among the earliest printed volumes. The title page of this one, published in England in 1526, offered readers "perfect knowledge" of herbs.

from seed to harvest in a single season, dying after the seeds form. Biennial herbs, which are also started from seed, produce foliage the first year; in the second year they bloom and the blossoms ripen into seeds; then the plants die. Annuals may be cultivated in all 50 states, but biennials need a cool period of rest between their first and second growing seasons, and those like caraway that are cultivated for seeds have difficulty surviving the warm moist winters in southern Florida and along the Gulf Coast; they grow well elsewhere. Many annuals and biennials will spring up in the garden year after year if the plants are allowed to produce and drop seeds; if you want to perpetuate such self-sown seedlings, you must learn to recognize them by their leaves and not pull them up as weeds.

To speed up the timetables for annuals and to produce plants that are ready for picking a month or more ahead of the garden schedule, start seeds indoors four to six weeks before the last frost is expected. Plant the seeds, three to a container, either in peat pots filled with packaged potting soil or in peat pellets *(drawings, page 26)*. Set them in a cool sunny window, and when the first true leaves appear above the pair of round seed leaves, cut off all but the strongest plant in each pot. Transplant the seedlings to the garden after all danger of frost is past, leaving them in their pots or pellets and spacing them as specified in the encyclopedia.

Perennial herbs live in the garden for many years—dying in fall and reappearing in spring. They can be grown almost anywhere in the country, although many gardeners find they do not do well in southern Florida and along the Gulf Coast. Lovage and mint do best in fertile soil, but most perennials prefer poor soil.

Except for tarragon, which produces sterile seeds, perennials can be started from seed and will become large enough to provide some usable leaves the first summer, though it is better to allow them to grow into their second year before harvesting many leaves. Perennials can be started from seed indoors in the same manner as annuals, but many gardeners, particularly those with limited window sill space, find it easier to sow the seeds outdoors, any time from early spring until early summer, in a well-drained seedbed or cold frame. Keep the soil slightly moist until the seedlings have developed a pair of true leaves above their seed leaves. Then thin them so they are 6 to 8 inches apart and leave them over the winter, transplanting them to the main garden the following spring.

If you do not need many plants and do not want to go to the trouble of sowing seeds, simply buy a few fully grown plants and increase them as needed by dividing the roots when the plants are dormant in early spring or by making stem cuttings any time in late spring to early summer. To divide the roots, simply dig up the plants with a spading fork and separate the roots with your fingers, discarding the aging center. Each division should contain at least one bud for a strong new stem. Replant these divisions.

Making stem cuttings is almost as easy. You can start as many as half a dozen in a 6-inch flowerpot that has been filled with a rooting medium such as perlite, vermiculite or equal parts of sand and peat moss or the coarse sand used by builders. Cut off 4- to 6-inch pieces from the tops of the stems, strip off the lower leaves, and then dip each stem into hormone rooting powder, available from garden supply centers. Insert the stems about an inch deep and pat the rooting medium firmly around them; then water lightly. Enclose the pot in a clear plastic bag to help keep the cuttings moist, using pencils or small sticks to prop the plastic up above the leaves; then place the pot in a bright, but not sunny, location. Within a few weeks the cuttings should develop roots and may start new top growth as well. When the new roots are about 1 inch long—pull one up to tell—move the cuttings to separate pots filled with packaged potting soil. No bag is necessary at this point, but plants should be watered regularly and protected from the noonday sun until they are sturdy enough to be transplanted to a pot to winter indoors. The following spring they may be set in the main garden.

Like other perennials, bulb-type herbs such as chives, garlic and shallots will live for years if not harvested. The bulbs may be grown anywhere in the country, even in a hot humid climate like that of the Gulf Coast. They prefer fertile soil and are commonly planted as bulb segments or, in the case of chives, as already rooted bulbs; plants started from seed require two years to mature.

The two commonly grown shrubby herbs are rosemary and thyme. Evergreen rosemary is a tender shrub that can be grown in the garden year round only where winter temperatures rarely fall below 10°. The nearly evergreen thyme, on the other hand, can survive winters as cold as 15° below zero if it is protected with a mulch of salt hay. Normal harvesting will take the place of pruning to control the size and shape of both herbs—and the stem cuttings may be used to propagate new plants.

In our garden, the first herb crop is made up of thinnings from the annuals like dill and anise—tender tiny morsels that Margaret proudly folds into omelets or sprinkles over hot buttered baby carrots. By the time these seedlings are about 6 inches high we can begin to pick sprigs from the tips of the stems regularly. Biennial parsley is ready to be cut as soon as it is 3 to 4 inches tall; perennials and shrubby herbs can be picked any time after their new leaves are fully formed. For the best flavor, cut herbs early in the morning immediately after the dew has evaporated and before the heat of the midday sun dissipates their fragrance. Picking herbs encourages new growth as long as you remember to spread the harvest

HERB VARIETY IN A JAR

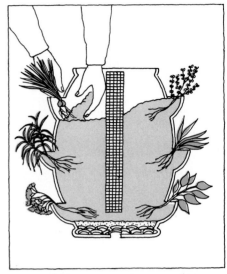

An old-fashioned strawberry jar provides a practical and attractive way to grow a collection of herbs on a sunny patio in summer, indoors in winter. The wide-topped jar can accommodate tall herbs, but the best candidates for the earlike planting pockets on the sides are small or trailing herbs like thyme or chives, preferably perennials that do not have to be replaced every season. To prepare the jar, line the bottom with a 2-inch layer of shards from a broken clay pot and 1 inch of coarse gravel; in the center stand a 3-inch cylinder of ¼-inch galvanized wire mesh filled with more gravel. Add packaged potting soil up to the lower edge of the bottom pockets. Insert a plant into each pocket, sliding it into place from inside the jar to keep the soil around the roots intact. Cover the roots with soil, fill the jar up to the next set of pockets and repeat. Keep the herbs moist by pouring water into the central drainage tube.

among several plants rather than cutting one back so severely that you deplete its energy. Removing blossoms as they appear will also force plants, especially annuals, to develop fresh foliage.

PRESERVING HERBS

Although the taste of fresh herbs cannot be matched, herbs preserved by drying or freezing come surprisingly close and, once processed, keep for a long time. Drying concentrates the oils that give flavor to the leaves and seeds and preserves both the taste and texture of garlic and shallot bulbs. In dried form, leaves and bulbs can be kept conveniently at room temperature for up to a year, seeds for three years or more. Freezer storage will keep leaves green and flavorful for up to a year but the herbs will wilt as soon as they are defrosted. When using them in cooking, frozen herbs may be substituted for an equal measure of fresh ones, but dried leaves or seeds are twice as pungent, so cut the measurement in half.

For drying or freezing, harvest the leaves just before the plants bloom. At that time the leaves are fully formed but still growing vigorously, so they possess the maximum amount of the essential oils that give them their distinctive flavors. Large leaves of herbs like sage or basil may be snipped off individually, but the easiest way to harvest the foliage of most herbs is to cut off the stems. Always keep at least two sets of leaves at the base of each stem so the plant will continue growing. Sometimes you can get two or three crops of leaves in a single season.

Delicate-tasting or tender-leaved herbs such as borage, chives and parsley lose flavor in the drying process and are better preserved by freezing; they should be washed, patted dry with paper towels and frozen whole or chopped in plastic bags. Pungent thick-leaved herbs such as sage, basil, summer savory and rosemary, on the other hand, keep best if dried. After washing the leaves, tie the stems in small bunches and hang them upside down in a dry, airy, shady place indoors for a day or two. When the leaves wilt and shrivel, slip each bunch into a paper bag punched with ventilation holes and large enough to cover the herbs without touching them and absorbing their oils, and hang them up to dry for 10 days to two weeks longer. When the leaves become crisp, rub them from the stems by kneading the sides of the bags. Then pour the mixture out into a bowl and remove and discard as many bits of stem as possible. Finally, spread the leaves on a cookie sheet and put them in an oven set to its lowest heat level and heat them until they are dry. Pack the dried leaves in airtight glass or metal containers.

DRYING SEEDS AND BULBS

Seeds, the tastiest parts of many herbs, are naturally wrapped in sturdy covers that will hold their flavor oils intact and usable for many years. If you plan to dry seeds of herbs like anise, caraway, co-

The chart at right lists the principal ways that herbs are used as flavorings in five major categories of foods. Herbs such as sage and chives are also added to cheese, anise and caraway to cookies and bread.

A cook's guide to 23 basic herbs

HERB	DRINKS AND SOUPS	MEATS	POULTRY AND EGGS	FISH AND SHELLFISH	VEGETABLES AND SALADS
ANISE	Whole or pulverized seeds in vegetable soups	Whole or pulverized seeds with sausages and roast pork			Whole or pulverized seeds in beet salads
BORAGE	Fresh flowers and leaves in cold fruit or wine punches	Chopped fresh leaves in sauces for meats	Chopped fresh leaves in omelets or deviled eggs	Fresh leaves in poaching liquid for fish and shrimp	Chopped fresh leaves in cabbage or green salads
BURNET	Chopped fresh leaves in vegetable soups and wine drinks	Chopped fresh leaves in chicken aspic		Chopped fresh leaves in fish and seafood sauces	Chopped fresh leaves in green salads or with celery or peas
CARAWAY		Pulverized seeds in meat loaf, goulash or sauerbraten	Pulverized seeds in filling for roast goose		Whole seeds in sauerkraut, potato salad or cauliflower
CHERVIL	Chopped fresh or dried leaves in vegetable soups	Chopped fresh leaves in sauces for lamb or beef	Chopped fresh leaves in omelets	Chopped fresh leaves on oysters	Chopped fresh leaves in potato, spinach or green salads
CHIVES	Chopped fresh leaves on cold soups	Chopped fresh leaves in beef and pork gravies	Chopped fresh leaves in omelets and chicken casseroles	Chopped fresh leaves in shrimp sauces and seafood stews	Chopped fresh leaves in green salads and on hot vegetables
CORIANDER	Whole seeds in hot fruit or wine punches	Pulverized seeds or leaves in chili con carne and curries	Pulverized seeds or chopped leaves in poultry stuffing		Pulverized seeds in fruit salads and with cold beets
DILL	Seeds or leaves in beet and tomato soups	Fresh or dried leaves in lamb stews or lamb sauces	Fresh or dried leaves in scrambled eggs	Fresh or dried leaves with salmon	Seeds or leaves with potatoes or cucumbers
FENNEL	Seeds or leaves in spinach and cabbage soups	Pulverized seeds or leaves in beef and lamb stews	Chopped fresh leaves in omelets	Chopped fresh leaves in fish stuffing	Chopped fresh leaves in stewed tomatoes or green salads
GARLIC	Chopped cloves in vegetable soups	Chopped cloves in meat sauces and marinades	Chopped cloves in chicken sauces or with braised chicken	Chopped cloves in fish stews and casseroles	Chopped cloves in salad dressings
LEMON BALM	Sprigs in iced tea; leaves for herb teas	Chopped leaves in marinades and sauces for lamb	Chopped fresh or dried leaves in poultry stuffings	Sprigs as garnish for fish; chopped leaves in fish stews	Chopped fresh leaves with asparagus
LOVAGE	Chopped fresh or dried leaves in bean soups	Seeds or leaves in meat pies and gravies	Chopped fresh or dried leaves in poultry stuffing	Seeds or leaves in poaching liquid for shrimp	Seeds or leaves in potato salads and slaws
MINT	Sprigs or leaves in juleps, iced tea or herb teas	Fresh leaves for mint sauces to accompany lamb		Sprigs or leaves as stuffing for fish	Chopped fresh leaves to garnish carrots, green peas or beans
PARSLEY	Chopped fresh leaves as garnish or ingredient for soup	Chopped fresh leaves in meat loaf, stews and casseroles	Chopped fresh leaves in omelets, poultry stuffing and sauces	Chopped fresh leaves in fish stews	Sprigs or chopped fresh leaves in green salads
ROSEMARY	Fresh or dried leaves in split-pea soup	Fresh or dried leaves with lamb or veal	Fresh or dried leaves in chicken sauces	Fresh or dried leaves in salmon stuffing	Fresh or dried leaves with turnips, broccoli or cabbage
SAGE	Fresh or dried leaves for herb tea	Chopped fresh or dried leaves with pork	Chopped fresh or dried leaves in poultry stuffing		Chopped fresh or dried leaves with beans or creamed onions
SHALLOTS	Chopped cloves in chicken, fish and vegetable soups	Chopped cloves in meat stews.	Chopped cloves in soufflés and with braised chicken	Chopped cloves in fish stews and sauces	Chopped cloves in salads and marinades
SUMMER SAVORY	Fresh or dried leaves in tomato or bean soups	Fresh or dried leaves in beef or lamb stews	Fresh or dried leaves in chicken fricassees	Fresh or dried leaves in fish stuffings	Fresh or dried leaves with green beans
SWEET BASIL	Chopped fresh or dried leaves in vegetable soups	Chopped fresh or dried leaves in meat sauces	Chopped fresh or dried leaves in poultry stuffing and stews	Chopped fresh or dried leaves in cream sauces for fish	Chopped fresh leaves for tomato salad or herb vinegars
SWEET CICELY	Chopped fresh leaves in beet or carrot soups	Chopped fresh leaves in beef or lamb stews	Chopped fresh leaves in omelets	Chopped fresh leaves in sauces for fish	Seeds or leaves in fruit or potato salads
SWEET MARJORAM	Chopped fresh or dried leaves in pea and potato soups	Chopped fresh or dried leaves in pork or veal sausages	Chopped fresh or dried leaves in chicken pies	Chopped fresh or dried leaves in creamed fish dishes	Chopped fresh or dried leaves with mushrooms
TARRAGON	Fresh or dried leaves in fish chowders	Fresh or dried leaves in veal stews	Fresh or dried leaves in omelets or chicken sauces	Fresh or dried leaves in fish and lobster sauces	Fresh or dried leaves in sauces or vinegars
THYME	Fresh or dried leaves in chowders or herb teas	Fresh or dried leaves with beef or lamb roasts	Fresh or dried leaves in poultry stuffings	Fresh or dried leaves in gumbos	Fresh or dried leaves in tomato aspic

riander, dill and lovage, keep an eye on the plants after they blossom. Seeds must be gathered when they are barely ripe—as soon as they begin to look brownish. In a day or two the seeds will begin to drop and the slightest jarring of the stems will send seeds flying in all directions. But because all of the seeds do not ripen simultaneously, not even those on a single plant, plan to pick several small crops over a period of a week or more. The best time of day to pick the seeds is in the early morning after the dew has evaporated but while the air is still relatively calm. Cut off entire seed heads with shears, dropping them into a paper bag as you go along. To get rid of the tiny insects that cling to seeds, drop the seed heads into boiling water for an instant. Skim off the dead insects, which will float to the surface, then drain the seeds. Spread them in a single layer on tightly woven cloth towels, then put the seeds in an airy shady place to dry for two weeks or until you can separate seeds from pods by rubbing them between your hands.

When the seeds have been loosened from the pods, pour them through a sieve or colander to separate them from the husks. To remove tiny bits of chaff, you may have to resort to winnowing. In this ancient process a current of air is used to carry the debris away. Choose a day with a light breeze and pour the seeds slowly from a height of 4 to 5 feet onto a sheet laid on the ground; the seeds, which are relatively heavy for their size, will fall to the sheet, but the chaff will blow away. When the seeds are completely cleaned and dried, pack them into airtight glass or metal containers.

Compared to drying herb seeds, drying garlic and shallot bulbs is simplicity itself. Wash the bulbs with a garden hose, then hang them by their braided stems to dry for a few days in an airy place out of the sun. If you dry bulbs in a garage or shed, be sure to bring them indoors before freezing weather and keep them in a well-ventilated cupboard over the winter to prevent rot.

HERB HOUSE PLANTS

Useful as dried herbs are, fresh ones are often better, and many gardeners are glad to pamper herbs in the house through the winter. Not all herbs do well indoors, but the list of possibilities is long enough to satisfy most cooks: sweet basil, chervil, chives, dill, fennel, lemon balm, mint, parsley, tarragon and thyme. Growing mature herbs indoors from seed is a slow process. For a fast crop, the types that must be started from seed—sweet basil, chervil, dill, fennel and sweet marjoram—should be sown in pots outside in late summer. Herbs can be grown in either clay or plastic pots, but good drainage is so important that I prefer to use unglazed clay pots even though their porous sides make watering more of a chore. Fill 3- to 4-inch pots with packaged potting soil and plant three seeds in each one, barely covering them with soil. Set the pots in a shady part of the garden and keep the soil moistened until the seeds germinate. Then move the pots into full sun and let the seedlings grow to about 1 inch in height before thinning out all but the largest one in each pot. Leave the pots outdoors until just before frost so that the plants will be growing as vigorously as possible when they are moved inside.

To grow lemon balm, mint, parsley, tarragon and thyme indoors, pot growing plants from the garden. Begin the move early by digging up the entire plant in late summer with a generous ball of soil around its roots. Set each plant into a 4- to 5-inch pot and cut back the top foliage by about half to balance the loss of roots damaged in the digging. Keep the plants in the shade of a tree for a week or so until the leaves stop wilting, then gradually give the plant more hours of direct sunlight each day. During this adjustment period, new roots will form and the herbs should be ready to be moved into the house early in the fall. Chives can be potted in the same manner but need a dormant period; they should be left outside in a sunny place for a month or so at freezing temperatures before they join the window garden.

Once inside, herbs need as much sun as you can give them. They grow best where night temperatures are low and the humidity is high. (Rosemary is an exception, requiring dryness.) During the day herbs can tolerate temperatures of 70° or higher, but they need cool nights for resting and ideally the temperatures should drop to 60° or lower. Choose an unheated room for the herbs, or keep the plants on a bedroom window sill, where night temperatures are likely to be lower than in the rest of the house. Most homes are as dry as deserts and unless you have a cool window sill in the kitchen, where the air is steamy from cooking, you will need to raise the humidity around the herbs. To do this, set them in a shallow tray lined with pebbles and keep the tray filled with water to a level safely below the bottom of the pots.

Whatever room you choose for your indoor herb garden, you will discover that the plants tend to be one half to one third of their garden size. Part of the reason for this is that the roots of the herbs are confined by the pots. But the main reason may be that you'll be pruning the foliage each time you snip off a sprig.

After a season in the house, the flavors of most herbs begin to dim. Margaret's plants become spindly no matter how many times she rotates the pots to equalize the sunlight on their foliage. The rosemary and thyme will revive, so we nurse them along until it is warm enough to return them to the garden, but the rest of the plants are rarely worth saving. Instead I turn the window sill over to new seedlings, which will provide a supply of tasty seasonings along with our fresh fruits and vegetables next year.

MAKING HERB VINEGAR

Herbs such as marjoram, sweet basil, caraway, dill, mint and tarragon can turn almost any vinegar into a distinctive treat for salad dressings, stews and sauces. The preparation is simple: Drop about a cupful of crushed fresh leaves into a 1-quart Mason jar that has been prewarmed, then bring the vinegar to a boil and pour it in. Cover the jar and set it aside for a week or two to let the flavor develop. If the herb taste is too strong, add more plain unheated vinegar; if the taste is too bland, add more fresh leaves and set the vinegar aside for another week. Finally, filter the liquid through a piece of muslin or toweling to remove any bits of leaves and keep it in a sealed glass jar or bottle, ready for use.

Treats for palate, nose and eye

To most people herbs mean seasonings. Yet the lift they give food is only one of the varied pleasures herbs offer. They can be used for dyes, ornaments for room decorations, scents, home-brewed teas, and in the landscape around the house, eye-catching delights of foliage color and pattern.

Many of these uses are illustrated on the following pages in the work of Mrs. Adelma Simmons, who grows nearly 400 different types of herbs on a farm in North Coventry, Connecticut. Starting with a single seedling of thyme some 40 years ago, Mrs. Simmons began to collect the kinds of plants that would have been grown in the herb patch when the farmhouse was built about 1790, and to experiment with using them in the ways popular then. Now Mrs. Simmons busies herself with herbs every month of the year.

In addition to culinary herbs—which she grows outdoors and in to make flavored vinegars and salts as well as to season soups and salads—Mrs. Simmons cultivates unusual varieties such as lady's-bedstraw, which she boils to produce a yellow dye; pennyroyal, which she dries and brews to make a mint-like tea; and camomile, from which she concocts a home remedy for headaches and sinus trouble. To create the decorative dried wreaths shown in the picture at right, Mrs. Simmons embellished rings of Silver King artemisia with pearly everlasting, strawflower, yarrow, statice and tansy. The heart-shaped sachet at the center of the picture, used to scent a bureau drawer, is perfumed with the aromatic herb rose geranium. The pomander ball at top left, whose odor is intended to discourage moths, is an orange pierced with whole cloves and rolled in a blend of dried spices and powdered orrisroot, an herb that helps keep the scent of the spices fresh.

Mrs. Simmons finds herb gardens fascinating to plan, and over the years she has designed and planted dozens of gardens for herself and other herb fanciers, ranging from elaborate formal beds like those of the purely decorative knot garden shown on pages 68 and 69 to the casual-looking but useful dooryard garden pictured on page 70.

A pomander, sachet and wreaths of dried aromatic and decorative herbs ornament a group of old herb books.

Hardy lavender (foreground) and broom (at right, behind it) mark the beds of a snow-covered herb garden at Caprilands Herb Farm in North Coventry, Connecticut.

Herb seedlings thrive in a cool sunny window. In the front, from left, are sweet marjoram, sweet basil and pennyroyal; in back are purple basil, pennyroyal and lavender.

Plants to make patterns

Although herbs are generally considered utilitarian plants, many are handsome and varied enough in color, texture and shape to be used as ornaments in the garden or on a window sill. Especially decorative and aromatic herbs—lavender and lamb's ears are two—bring scent and visual delight to gardens like the formal one shown overleaf. Even in winter, perennial herbs *(below, left)* continue their landscaping role, preserving the outline of garden beds, while seeds are started indoors *(below)* to provide new plants for spring.

The colorful foliage of lavender and potentilla (circular bed, center), catnip (triangles) and silvery lamb's ears (half moons) decorates

an ornamental herb garden, its knot pattern based on a French tapestry design. Creeping thyme covers the walks between the beds.

Herbs for scents and savors

Whether culinary herbs are used to flavor summer salads or dried for later use *(overleaf)*, they taste best if they are harvested in their prime. One practical way to make sure that tender young leaves or ripe seeds will not be overlooked is to locate the herbs in vegetable beds *(below, right)*, where the crops may be planted and picked together. Another answer, which takes advantage of the plants' aromatic virtues, is to grow the herbs close to the kitchen door *(below, left)*, where their perfumes act as a constant reminder to the cook.

On a dooryard terrace, sage and thyme (center) are set between paving stones—where they can be walked on and crushed to release more of their pleasant scent.

Neat rows of comfrey (foreground), as well as sorrel, bush basil, chives and dill, thrive next to tomatoes in a vegetable garden, furnishing a steady supply of herbs for cooking.

To preserve the crop for winter use, Mrs. Simmons ties culinary, decorative and aromatic herbs into bunches and hangs them upside

down to dry in an airy shed. After drying, ornamental ones like the yellow strawflowers in the center foreground are stored in baskets.

A harvest of many uses

Months, even years, after herbs have been picked they go on giving pleasure in dried form. Culinary kinds, like those at right below, keep their flavor when they are packed in airtight containers. Many decorative herbs remain colorful and shapely; the Silver King artemisia shown at left below can be bent around wire rings to form dried wreaths that make handsome gifts. And herbs like lavender, which still smells sweet after drying, can perfume a room in the aromatic mixture of roots, leaves and flowers called potpourri *(overleaf)*.

Wreaths of artemisia (foreground) and statice (behind it) will remain decorative long after the nutmegs and cinnamon sticks that adorn them have been plucked off for cooking.

During the months when fresh herbs are out of season, a collection of dried culinary herbs keeps the cook stocked with a welcome variety of flavorings for everyday dishes.

Aromatic dried herbs such as lavender (bottom) and rosebuds (top) are blended with spices and ground orrisroot (left) to produce

the old-fashioned potpourri at right, which will be preserved in a jar that can be opened from time to time to scent the air indoors.

An encyclopedia of vegetables, fruits, nuts and herbs 5

To help you select and grow the varieties best suited to your purposes, the following encyclopedia provides planting, cultivating and havesting information for 78 of the most popular home-grown crops: 34 vegetables, 16 fruits, 5 nuts and 23 culinary herbs. The entries also indicate where and when various plants grow best and the average quantities of produce they may be expected to yield.

Unless otherwise noted, all the vegetables and herbs listed can be grown anywhere in the United States and southern Canada. Some vegetables, such as broccoli, require growing seasons with certain minimum and maximum temperatures; local weather bureau statistics can help you determine if and when the necessary conditions prevail where you live. Other crops, such as corn, are planted in relation to the date of the last expected frost in spring or the first expected frost in fall; maps showing the approximate dates of these frosts appear on page 149. The climate zones in which different fruits and nuts do best are designated by numbers keyed to the map on page 148.

All the plants listed grow best in full sun, and unless otherwise noted, require moist well-drained soil of average fertility. Most herbs need an especially well-drained soil that is not too rich; they should not be fertilized unless the entry so indicates. Most vegetables, on the other hand, need fertilizer to encourage rapid growth and succulence; instructions appear in each vegetable entry. Fruits and nut trees should not be fertilized when they are planted; if fertilizer is needed after growth is established, instructions are given. The identification tag that is attached to a branch of all purchased fruit trees should be removed promptly to avoid constricting growth and killing the branch. To eliminate weeds and conserve moisture, it is a good idea to mulch your plants; for control of pests and diseases see the charts on pages 150-154. Except where noted, vegetable seeds retain their vitality for three to four years, herb seeds for two to three years; unused seeds should be stored in the refrigerator in an airtight container for sowing in subsequent years.

The bounty of a home garden is illustrated by this group of herbs, vegetables, fruits and nuts, which includes feathery dill (upper left), red peppers (upper right), blueberries (in basket) and almonds (bottom right).

Vegetables

A

ASPARAGUS
Asparagus officinalis

Unlike most other vegetables, which are annuals, asparagus is a perennial plant that may remain productive for 20 years or more. However, it requires a dormant period during winter months and grows only where dormancy is induced naturally by climate or can be forced artificially by cultivation methods. It grows well anywhere in the United States and southern Canada except Florida and along the Gulf Coast, where the moist soil and mild temperatures prevent it from getting its necessary period of rest.

Asparagus does not produce the first crop until its third season if it is grown from seeds, or until its second season if it is started from year-old roots. Thereafter, it is one of the first vegetables to be ready for picking each spring. The plant has fine fernlike foliage and would grow 4 to 6 feet tall if left unpicked, but it should be harvested when its stems are only about half a foot tall. One of the best varieties for most areas is Waltham Washington. A widely used improved form in the Midwest is Viking, sometimes called Mary Washington Improved. Of special value in California is a variety called 500W. A 12-foot row yields about 6 pounds of asparagus over a period of 6 to 10 weeks.

HOW TO GROW. Asparagus grows best in soil with a pH of 6.0 to 8.0. Sow seeds indoors in midwinter or in a hotbed in early spring, soaking them first in tepid water overnight, then planting them 1 inch apart and ½ inch deep. When all danger of frost is past, transplant the seedlings to a nursery bed for one season; space them 4 inches apart. The next spring as soon as the soil can be worked, transplant the seedlings to the garden, preferably off to one side where they will not interfere with annual crops. Space them 1 foot apart in rows 1½ to 2 feet apart.

To save a year's growingtime, most home gardeners start asparagus from strong one-year-old roots bought in early spring and set directly into the garden, as shown in the drawings on page 22. Two to three months after planting roots, scatter a 1-foot-wide band of 5-10-5 fertilizer along each side of the row at the rate of ½ pound per 10 feet of row. Early each spring, beginning with the second season, apply 1 to 2 pounds of 5-10-5 fertilizer to each 10 feet of row before growth starts. To induce dormancy in the warm, arid Southwest, do not irrigate in winter.

Asparagus should be picked when the stems reach 8 inches in height and the buds at the tips of the stems are still tightly compressed. Pick all stems at this stage; if any stems are allowed to grow larger, the plants' ability to send up new stems will be greatly diminished. The picking season is over when stems no longer grow larger than ½ inch in diameter. To pick asparagus, bend the stems until they snap; the portion that is too tough to snap is also too tough to eat. Do not use an asparagus knife, a special V-shaped device that commercial asparagus growers employ to cut the plant underground, for it leaves tough white tissue at the base of the plant. This white tissue, together with the lower part of the green stem, must always be discarded in the kitchen because the fibers are too stringy to eat. (Store-bought asparagus includes these waste parts.) Cutting the plant also often injures other stems underground before they have a chance to emerge.

ASPARAGUS

B

BEAN
Snap bean, also called string bean, green bean or wax bean; shell bean, also called horticultural bean *(Phaseolus vul-*

garis); Lima bean, also called butter bean *(P. limensis,* also called *P. lunatus macrocarpus);* baby Lima bean, also called baby butter bean *(P. lunatus);* edible soybean *(Glycine soja,* also called *G. max)*

Beans are among the most important food crops, economically and nutritionally, in the world. The pods of all the species listed here contain beans that can be cooked or dried for later use; in the case of snap beans, the pods themselves are cooked. The plants bear tiny white, yellow, pink, red or lavender flowers that resemble sweet-pea blossoms.

Snap beans and shell beans may be grown in all parts of the U.S. and southern Canada. Lima beans, baby Lima beans and edible soybeans do best where summers are long and hot, and night temperatures remain above 50° for a period of at least two months (four months in the case of edible soybeans).

The most valuable bean for the home gardener is the snap bean. (The alternate name string bean is no longer accurate, because modern types are stringless; the term refers to the tough fibers that joined the two halves of the pods of older varieties.) Snap beans come in bush varieties, which usually grow about 1½ feet tall, and pole, or climbing, varieties, which twine counterclockwise around any support and become 6 feet or more tall. The plants produce green, yellow and, occasionally, purple pods. Among bush-type snap beans, excellent green-podded varieties are Executive, Improved Tendergreen, Tendercrop, Topcrop and Bush Romano (the latter has tender, broad pods and the meaty flavor usually found only in plants known as Italian pole beans). Fine bush types of wax, or yellow-podded, snap beans are Brittle Wax, Cherokee Wax, Kinghorn Wax, Pencil Pod Black Wax and Goldcrop. A variety of purple-podded bush that is green when cooked is Royalty. A 15-foot row of bush beans yields about 7 pounds over a period of two to three weeks. Among pole beans, the most popular green-podded type is Kentucky Wonder, but Blue Lake and McCaslan are also fine; a top-quality Italian pole bean is Romano. A 15-foot row of pole beans yields about 12 pounds over a period of six to eight weeks.

In addition to snap beans, the species *Phaseolus vulgaris* includes a special group of beans known as shell, or horticultural, beans, which have a mealy texture and delicious nutlike flavor. Three excellent bush varieties of shell beans are Seneca Horticultural, French Horticultural and Dwarf Horticultural, also called Wren's Egg, all of which grow about 18 inches tall. A 15-foot row of bush shell beans yields about 9 pounds of beans over a period of three weeks. Speckled Cranberry, also called Horticultural, is a fine pole-type shell bean that grows 6 to 8 feet tall. A 15-foot row of pole shell beans yields about 18 pounds of beans over a period of four weeks.

Lima beans and baby Lima beans, grown for their ½- to ¾-inch immature beans, have a dry, mealy texture and nutlike flavor, and are exceptionally nutritious. Bush Lima beans grow about 2 feet tall; excellent varieties are Fordhook 242 and Burpee's Improved Bush Lima. Pole Lima beans grow 8 feet or more tall; fine varieties are King of the Garden, Burpee's Best and Prizetaker. Bush varieties of baby Lima beans grow 2 feet tall; good varieties are Baby Fordhook Bush Lima and Thorogreen. Pole baby Lima beans grow 8 feet or more tall; recommended varieties are Carolina, also called Sieva, and Florida Butter. A 15-foot row of bush Lima beans or baby Lima beans yields about 4 pounds of beans over a period of three to four weeks. Pole varieties yield about 7 pounds over the same period of time.

Edible soybeans grow 2 to 2½ feet tall. The beans are usually cooked when they are green, but if the pods are al-

SNAP BEAN
wax and green types

For climate zones and frost dates, see maps, pages 148-149.

lowed to mature and turn yellow, the beans can be cooked like Lima beans. Two good varieties are Bansei and Kanrich. A 15-foot row of soybeans yields 8 to 12 pounds of beans over a period of two weeks.

HOW TO GROW. Beans grow best in soil with a pH of 6.0 to 7.5. To reduce the risk of disease, do not plant them where any other beans have grown in the past three years.

Beans are legumes—that is, they draw the nitrogen they need for growth from the atmosphere with the help of bacteria in the soil. To ensure that enough bacteria are present, bean seeds should be dusted with a commercial preparation of bacteria, available from garden supply centers, before they are planted.

In most of the U.S. and southern Canada, where frost is expected in winter, sow seeds of bush varieties of snap beans and shell beans about the date the last spring frost is due, and make additional plantings every two weeks until about eight weeks before the first fall frost is due; this succession of plantings will assure fresh beans all summer. Sow seeds of pole varieties of snap beans and shell beans only once, when the last spring frost is due—the plants will continue to bear until frost if all pods are picked.

In frost-free regions sow seeds of Lima beans, baby Lima beans and edible soybeans in early spring when night temperatures no longer can be expected to drop below 50°. Continue to sow bush varieties of Lima beans and baby Lima beans until about 10 weeks before night temperatures are again expected to drop below 50°, pole varieties until about 13 weeks before 50° nights, and edible soybeans until about 15 weeks before 50° nights. In these frost-free regions sow seeds of all other types of beans in fall as soon as maximum temperatures begin to average below 80°. Continue to plant bush varieties of snap beans and shell beans until about two months before maximum temperatures are again expected to average 80° or higher, pole varieties of snap beans and shell beans until about three months before the arrival of 80° days.

Sow seeds 1 to 1½ inches deep where the plants are to grow, since beans are particularly difficult to transplant. For bush beans, set seeds singly about 3 inches apart in rows about 2 feet apart. Good poles for pole beans to climb on are freshly cut saplings with the bark still on; they can be set to stand singly and upright, or angled in groups tepee fashion for stability. Set poles about 2 feet apart in rows 3 feet apart, and plant four to six seeds around each; after the plants have sprouted, thin out all but the best three or four in each group. If pole beans are grown along fences, sow seeds singly about 6 inches apart.

When both bush and pole plants are about 6 inches tall, sprinkle a 6-inch band of 5-10-5 fertilizer along each side of the row at a rate of 5 to 8 ounces to every 10 feet of row. Keep the fertilizer off the leaves and 3 inches away from the stems of the plants. Avoid overhead watering, and never touch the plants when the leaves are wet.

Snap beans should be harvested while they are still immature—that is, while the pods are still tender, moist and succulent, and still able to snap when they are bent. All other types of beans are ready for harvesting when the beans inside the pods are fully formed (open a pod to see). The length of time required from seeding to harvest varies according to the type of bean planted: bush varieties of snap beans and shell beans need about 8 weeks; pole varieties of snap beans and shell beans, 9 weeks; bush Lima beans and bush baby Lima beans, 9 to 10 weeks; pole Lima beans and pole baby Lima beans, 13 weeks; and edible soybeans, 15 weeks.

If you have an oversupply of beans, you can dry them for future use by letting the pods mature on the vine until

SHELL BEAN

LIMA BEAN

they become beige-colored. Then remove the beans from the pods and heat them in a 130° to 145° oven for an hour to kill any bean weevils that may have burrowed into the pods while they were still green.

BEET
Beta vulgaris

Beets are cool-weather biennials that are grown as annuals, producing their globular or tapering roots and their reddish green tops—both used as cooked vegetables—during their first season. Garden beets have red, yellow or white roots. Excellent red varieties include Crosby's Egyptian, Early Wonder and Ruby Queen, early types that are ready to be harvested 55 to 60 days after seeds are sown; and Detroit Dark Red or King Red, which mature in 65 to 80 days and can be used for winter storage. Burpee's Golden, a yellow-rooted variety, and Burpee's White, which has white roots, mature in 55 to 60 days. Twenty-five feet of row yields about 20 pounds of beets over a period of two to three weeks.

HOW TO GROW. Beets grow best in soil with a pH of 6.0 to 7.5. In most of the U.S. and southern Canada, where frost is expected in winter, sow seeds in early spring as soon as the soil can be worked and continue planting at three-week intervals until 60 days before maximum daytime temperatures are expected to average about 80°. In late summer, when maximum daytime temperatures average below 80°, start successive plantings until 10 to 12 weeks before minimum night temperatures average below 20°. Beets are not harmed by spring or fall frosts. However, roots become tough in hot weather, so in regions where maximum summer daytime temperatures average above 80° for prolonged periods and winter night temperatures rarely fall below 30°, start successive plantings in early fall for harvesting during the winter and spring; make the final planting 60 days before maximum daytime temperatures are expected to average about 80°.

Beet seeds come in clumps containing three or more seeds. Sow the clumps ½ inch deep and 1 inch apart in rows 14 to 16 inches apart. Several plants will come up close together from each clump. Pull up all but the strongest plant in each group; the uncrowded survivor will form a fast-growing tender root. When the seedlings become 3 to 4 inches tall, scatter a 6-inch-wide band of 5-10-5 fertilizer along each side of the row at a rate of 5 ounces per 10 feet of row. During dry weather, water enough to keep the soil moist and prevent the plants from wilting. When the seedlings become 6 inches tall, pull up every other plant and use its greens and small root. Beets are ready to be pulled up for their roots eight to nine weeks after seeds are sown; the roots are most tender when they are less than 2 inches across. To see if the plants are ready, push away some soil and look at the roots before pulling them up. If the roots develop black bitter areas, the soil lacks boron. Correct this deficiency by mixing ¼ teaspoon of household borax in 12 gallons of water and sprinkling the solution lightly around the plants. Subsequent rains or watering will wash the boron into the root zone.

BROCCOLI
Brassica oleracea italica

Broccoli, a cool-weather relative of the cabbage, is grown for its clusters of flower buds and the tender stems and leaves near them. Plants become 2 to 3 feet tall with an equal spread. The tip of the main stem of each plant develops the largest cluster of buds. This terminal cluster is usually 6 inches in diameter, but on some varieties it becomes as much as 14 inches across. Green Comet Hybrid

BEET

For climate zones and frost dates, see maps, pages 148-149.

83

and Spartan Early are early varieties; Calabrese, Italian Green Sprouting and Waltham 29 are midseason varieties. A 15-foot row yields about 10 pounds of broccoli over a period of five weeks.

HOW TO GROW. Broccoli grows best in soil with a pH of 6.0 to 7.0. To reduce the risk of cabbage diseases, do not plant broccoli where other broccoli plants, cabbages or any other relatives of cabbage have grown within the past three years. The plants do best when temperatures remain between 40° and 70° over a growing period of 105 to 115 days. In most regions where winter frosts are expected, start early varieties of broccoli indoors or in a hotbed two to three months before the last spring frost is due, setting the seeds ½ inch deep. When the seedlings are 1 to 1½ inches tall, transplant them to individual pots and place them in a cold frame until about two weeks before the last frost is expected. Then set them into the garden, spacing them 1½ feet apart in rows 3 feet apart. Plant midseason varieties directly in the garden in early summer; seedlings tolerate temperatures over 70°, and the weather will have cooled by the time the plants begin to approach maturity and become sensitive to heat. In regions where winter temperatures rarely fall below 25°, sow seeds directly in the garden in early fall.

To start plants directly in the garden, group three or four seeds in a spot, setting each group ½ inch deep and 1½ feet apart in rows 3 feet apart. When the seedlings become 1 inch tall, pull out all but the strongest plant in each group. To prevent cutworm injury, slip over each seedling a paper cup that has had its bottom removed. Water enough to keep the soil moist and prevent the plants from wilting. Fertilize three times during the growing season —when the plants are 6 to 8 inches tall, when they are 12 to 15 inches tall and when the buds begin to form; scatter an 18-inch band of 5-10-10 fertilizer along each side of the row at the rate of 5 ounces of fertilizer per 10 feet of row.

The plants are ready for picking three and one half to five months after seeds are sown, just before the buds begin to open. Harvest broccoli with a knife, cutting the stems about 6 inches beneath the clusters of buds. After the main cluster is picked, side branches continue to produce smaller clusters of buds over a period of 8 to 10 weeks.

BRUSSELS SPROUTS
Brassica oleracea gemmifera

Brussels sprouts, cool-weather relatives of the cabbage, grow well where summer daytime temperatures average about 65° or less, especially along the east and west coasts of the U.S. and southern Canada; they do not grow satisfactorily in the high summer temperatures of Florida, along the Gulf Coast and in much of the Midwest. Each plant becomes about 2½ feet tall, bearing along its stem as many as a hundred 1- to 2-inch ball-like sprouts resembling miniature cabbages. Brussels sprouts are very resistant to cold, and since the tastiest sprouts are those that mature after the first fall frost, many gardeners time their planting so that picking can begin at the first expected frost. A good variety is Jade Cross Hybrid, well suited to all regions. Twenty-five feet of row yields about 8 pounds of sprouts over a period of six weeks or more.

HOW TO GROW. Brussels sprouts grow best in soil with a pH of 6.0 to 7.5. To reduce the risk of cabbage diseases, seeds should not be planted where other Brussels sprouts, cabbages or any other relatives of cabbage have grown within the past three years. Sow seeds about 120 days before the date of the first expected fall frost. Group three or four seeds in a spot, setting each group ¼ inch deep and 18 inches apart in rows 3 feet apart. When the seedlings be-

BROCCOLI

come 1 inch tall, pull out all but the strongest plant in each group. Protect young plants from cutworms by slipping over them paper cups that have had the bottoms removed. Fertilize three times during the growing season —when the plants are 6 to 8 inches tall, when they are 12 to 15 inches tall and again just as the sprouts start to form; scatter an 18-inch band of 5-10-10 fertilizer along each side of the row at the rate of 5 ounces of fertilizer per 10 feet of row. When the lower leaves begin to yellow, remove them to give the sprouts room to develop. To speed the development of the sprouts, pinch off the tip of each plant in early fall; this trimming results in earlier harvests but reduces the yield by about a third. The first sprouts are ready for picking about four months after seeds are sown, when they are firm; the rest continue to mature over a period of six weeks or longer. If night temperatures start dropping to about 20° on a regular basis before all the sprouts mature, dig up the plants with a little soil around them and put them in a deep cold frame or in an unheated garage; they will then probably continue to grow until all the sprouts mature.

BRUSSELS SPROUTS

C

CABBAGE
Brassica oleracea capitata

All cabbages are globular and are grown for their tightly compressed leaves, but there are three main types—green-foliaged cabbage, with smooth green leaves; red cabbage, with purplish red leaves; and savoy cabbage, with crinkly green leaves. They come in early, midseason and late varieties. Among the green-foliaged cabbages, recommended early varieties are Golden Acre, Jersey Wakefield, Marion Market and Stonehead Hybrid; midseason varieties are King Cole Hybrid, O-S Hybrid and Greenback; and late varieties are Seneca Danish Ballhead, Penn State Ballhead and Wisconsin Hollander. Fine red cabbages include Ruby Ball, early; Red Acre, midseason; and Red Danish Ballhead, late. Excellent savoy varieties are Savoy King and Chieftain, both midseason. Many seedsmen sell packages that combine the seeds of early, midseason and late varieties for gardeners who do not want to buy a package of each type. All need cool temperatures. Most early varieties weigh about 3 pounds; midseason and late varieties weigh 4 to 6 pounds. A 9-foot row yields about 24 pounds of cabbage over a period of three weeks.

HOW TO GROW. Cabbages grow best in soil with a pH of 6.0 to 7.5. To reduce the risk of cabbage diseases, the plants should not be placed where other cabbages or any relatives of the cabbage have grown within the past three years. In most of the U.S. and southern Canada, where winter frosts are expected, start early varieties indoors or in a hotbed about six weeks before the last spring frost is expected; set the seeds ½ inch deep. When the seedlings are 1 inch tall, transplant them to separate pots; then set them into the garden two or three weeks before the last frost is due. Midseason varieties can be sown directly outdoors about the time the last frosts are due, and late varieties about a month later except in the Midwest, where only early varieties are recommended because of the hot summers. In regions where winter temperatures rarely fall below 30°, sow seeds directly outdoors in late summer.

When sowing seeds in the garden, group three or four in a spot, setting each group ¼ inch deep. Space groups of early cabbage seeds about a foot apart in rows 2 to 2½ feet apart. Midseason and late varieties should be set 1½ feet apart in rows 3 feet apart. When the seedlings become 1 inch tall, cut off all but the strongest plant in each group. Protect seedlings from cutworm injury by

CABBAGE

For climate zones and frost dates, see maps, pages 148-149.

slipping over them paper cups that have had the bottoms removed. Fertilize at three- to four-week intervals, scattering a 12-inch band of 10-10-10 fertilizer along each side of the row at a rate of 8 ounces of fertilizer for every 10 feet of row. Keep the soil uniformly moist to prevent the cabbageheads from splitting. Cabbage roots lie very near the surface of the soil and are easily damaged by cultivation. If it becomes necessary to cultivate, penetrate the soil no deeper than 1 inch.

Pick cabbages when the heads are firm by cutting them off at the base of the stalk. Early varieties reach maturity 105 to 115 days after seeds are sown, midseason varieties take 125 to 135 days and late varieties 145 to 165 days.

CABBAGE, CELERY See Chinese Cabbage
CABBAGE, CHINESE See Chinese Cabbage
CABBAGE, TREE See Collard
CANTALOUPE See Melon

CARROT
Daucus carota sativa

The carrot, surprisingly, is a descendant of that delicate wild flower of the fields, Queen Anne's lace—if left unpicked in warm areas, it will in its second year produce tiny flat-topped white blossoms like its weedy ancestor's. As a vegetable, however, the carrot is grown not as a biennial but as a warm-weather annual, sending up a mound of bright green feathery foliage 10 to 12 inches tall as it develops its familiar orange-yellow roots. The type of carrot to grow depends on your soil: if you have light, sandy, stoneless soil 10 to 12 inches deep, you can grow varieties that have slender roots 8 to 9 inches long, such as Gold Pak, Imperator and Waltham Hicolor; if your soil is shallow or rocky or consists of heavy clay, select varieties with roots 6 to 7 inches long, such as Chantenay, Nantes, Danvers 126 and Danvers Half Long, or else use those with thick roots 4 to 5 inches long, such as Burpee's Oxheart and Short 'n Sweet. A 15-foot row yields about 8 pounds of carrots over a period of three weeks.

HOW TO GROW. Carrots grow best in light, sandy soil with a pH of 6.0 to 7.0. Since they are most tasty when young, it is a good idea to sow successive crops at three-week intervals. They grow best when temperatures are between 40° and 80°, so in most of the United States and southern Canada, where winter frost can be expected, begin to sow the first seeds in early spring as soon as the soil can be worked and continue to sow at three-week intervals until three months before minimum night temperatures average below 20°. Carrots will not grow well when maximum temperatures average over 88°. In regions where winter temperatures rarely fall below 25°, start sowing in late summer and continue to sow at three-week intervals until three months before maximum temperatures are expected to average 88°.

Space the seeds about four to an inch, setting them ½ inch deep in rows 14 to 16 inches apart. Sow a few seeds of radish (*page 104*) in with the carrots; radishes sprout quickly and will mark the row until the carrots appear. Carrot seedlings are extremely tiny. To prevent the soil from forming a crust that would inhibit the seedlings from breaking through the surface of the soil, cover the seeds with a light layer of vermiculite, sifted compost or grass clippings; firm the covering well and water the row. As the seedlings appear, thin them to stand 1 inch apart. When the carrots become ½ inch thick—dig away some dirt to see—pull every other plant, making the final spacing 2 inches apart. Do not discard the pulled plants; though thin, they are edible and delicious at this stage.

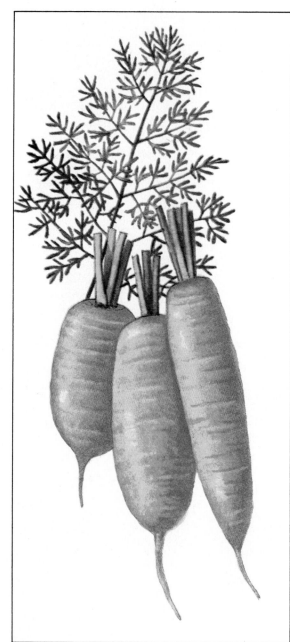

CARROT
short, medium and long types

Fertilize twice—when the plants are 3 to 4 inches tall and when they are 6 to 8 inches tall—scattering a 6- to 8-inch band of 5-10-10 fertilizer along each side of the row at the rate of 5 ounces per 10 feet of row. After each feeding, pile about ½ inch of soil around the base of each plant to prevent the tops of the roots from turning green from exposure to light. Carrots become fully mature 65 to 85 days after seeds are sown but make good eating earlier.

CAULIFLOWER
Brassica oleracea botrytis

Cauliflower is a cabbage relative grown for its flower buds, which are clustered together in a head (sometimes called a curd). The plant is somewhat difficult to grow, requiring cool temperatures, constant moisture and frequent fertilizing. Cauliflower grows about 2½ feet tall and has long blue-green leaves.

There are two main kinds of cauliflower—one kind has white buds, the other has purple buds (the purple buds turn green when cooked). To ensure the production of firm as well as white heads, the white varieties must be blanched—that is, their buds must be shielded from light, a procedure that changes their color from green to white. Purple-budded varieties form solid heads naturally and do not require blanching. Typical good white varieties are Snow King Hybrid, Burpeeana and Seneca Snowball. Purple Head is a good purple-budded variety. Some seedsmen sell seed packages in which both types are mixed. A 7-foot row will yield about five heads over a period of five weeks.

HOW TO GROW. Cauliflower grows best in soil with a pH of 6.0 to 7.5. To reduce the risk of cabbage diseases, seeds should not be planted where other cauliflowers, cabbages or any other cabbage relatives have grown within the past three years. In regions where frost is expected in winter and maximum summer daytime temperatures average 75° or less, cauliflower can be grown as a spring or fall crop. In regions where winter temperatures rarely fall below 30°, it can be grown as a winter crop.

For a spring crop, sow seeds indoors or in a hotbed four to six weeks before the last spring frost is expected, setting the seeds ½ inch deep. When the seedlings are 1 to 2 inches tall, transplant them to individual pots. Set the plants into the garden at about the time of the last frost (light frost does not harm them); space them 18 to 24 inches apart in rows about 3 feet apart. For a fall crop, sow seeds directly into the garden in late spring. For a winter crop, sow seeds outdoors in late summer.

To sow seeds in the garden, group three or four seeds in a spot, setting each group ½ inch deep and 18 to 24 inches apart in rows about 3 feet apart. When the plants are 1 inch tall, pull out all but the strongest plant in each group. Because of the wide space between cauliflower plants, lettuce, radishes or other quick-maturing crops can be planted between rows and in the same row.

Protect young cauliflower plants from cutworms by slipping over them paper cups that have had the bottoms removed. Keep the soil constantly moist, and fertilize every three to four weeks, scattering a 12-inch band of 10-10-10 fertilizer along each side of the row at the rate of 5 to 8 ounces per 10 feet of row. If cultivation is necessary, do not dig deep because cauliflower roots lie close to the surface of the soil and are easily injured. Blanch white-budded cauliflower on a dry day when the clusters are about 2 inches across. To blanch the heads, pull a few outer leaves together over the buds. Gather the leaves into a topknot and hold them together loosely with a string or rubber band. In warm weather the buds may take two to four days to turn white; in cold weather they may take a week or more.

CAULIFLOWER

For climate zones and frost dates, see maps, pages 148-149.

CELERY

CELERIAC

White-budded varieties are ready to be picked 100 to 110 days after sowing, purple-budded plants in 130 to 145 days. Harvest the buds while they are very tight by cutting the stalk just below the head; quality deteriorates if the buds begin to open. If night temperatures start to fall below 25° for more than a few nights at a time before the plants mature, dig them up with a little soil around them and put them close together in a deep cold frame or in an unheated garage; they will probably mature in a few weeks.

CELERY and CELERIAC

Celery *(Apium graveolens)*; celeriac, also called turnip-rooted celery and knob celery *(A. graveolens rapaceum)*

Celery and celeriac are not as widely grown in home gardens as other vegetables because they require a cool growing season of five or six months' duration, doing best when average daytime temperatures stay between 60° and 70°. Such weather occurs in widely separated regions at different times of the year—in summer in southern Canada, in the northern parts of the United States, particularly around the Great Lakes, and in coastal areas, such as Long Island and the Pacific Northwest; and from late fall until spring in the South and in California. Both plants grow 15 to 18 inches tall.

Celery is grown primarily for its delicate-tasting leafstalks, but the leaves are also edible. Celery comes with either green stalks or yellow stalks. Years ago green-stalked varieties, especially those grown commercially, were often blanched—that is, the stalks were shielded from the light so that they would lose their green coloration. This practice has generally been abandoned because of the labor involved and because green celery has many more vitamins than blanched celery. Types that naturally have yellow stalks are bred to that color and are called self-blanching. Fine green selections are Summer Pascal, an early variety that is also known as Waltham Improved; and Burpee's Fordhook, Giant Pascal and Utah 52-70, all late varieties. A fine early yellow variety is Golden Self-Blanching. A 15-foot row of celery plants will yield approximately 30 stalks during a period of five weeks.

Celeriac is grown for its globular root, which has a celerylike flavor and usually is about 4 inches in diameter when it reaches maturity. Unlike celery, it is usually eaten cooked rather than raw. A good variety is Alabaster. A 10-foot row of celeriac produces about 6 pounds over a period of eight weeks.

HOW TO GROW. Celery and celeriac need an extremely rich moist soil with a pH of 6.0 to 7.0. Most home gardeners who grow celery buy 4- to 6-inch-tall seedlings; they can be planted when night temperatures are no longer likely to fall below 40° (lower temperatures make the plants send up inedible flower stalks instead of edible leafy stalks). Prepare the ground a month before planting, making a bed 12 to 18 inches wide consisting of a 5- to 6-inch layer of compost or well-rotted cow manure; supplement the layer with 1 pound of 10-10-10 fertilizer for every 10 feet of row. Space the beds about 2½ feet apart. Set the plants in a double row in each bed, spacing them 6 to 8 inches apart in both directions. Keep the plants constantly moist. Fertilize every two to three weeks during the growing season by pouring around the plants 1 pint of water-soluble liquid fertilizer diluted to one half the strength recommended on the label.

Celery may also be started indoors from seeds. In regions where winter frosts can be expected, sow the seeds in spring about 10 weeks before 50° to 60° night temperatures are due; in frost-free regions, sow seeds in early fall. Keep the plants on a warm sunny window sill. When the seedlings become about 1 inch tall, transplant them to individual 2½- to 3-inch peat pots. When the young plants become 3 to 4 inches tall, set them, pots and all, into the garden. Celery will generally reach its mature size approximately 110 days after the plants are set into the garden, or approximately 180 days after seeds are sown; however, they are edible at all stages.

If you wish to take the trouble to blanch celery, cover the stalks—but not the leaves at the tips of the stalks—in late fall with ordinary soil; or else lay 12-inch boards or sheets of plastic against each side of the plants, holding the shields in place with soil. Do not pile soil against the stalks in warm weather because it may cause the plants to rot. To harvest celery, pull up the plants and cut off the roots just below the base of each stalk.

Sow celeriac seeds indoors in regions where frost is expected in winter and outdoors in frost-free regions about 10 weeks before night temperatures can be depended upon to stay above 50°. Follow the planting procedures outlined above for celery. Celeriac matures approximately 200 days after seeds are sown. However, the root is edible at any earlier stage. To harvest celeriac, pull up the plants and cut off the tops; the tender roots may be stored in damp sand for several weeks.

CELERY CABBAGE See Chinese Cabbage

CHARD, SWISS
Beta vulgaris cicla

Swiss chard is a relative of the beet, but unlike the beet it is grown only for its tender, vitamin-rich leaves rather than for both the roots and leaves. The plants generally grow 1 to 1½ feet tall. The red or green deeply crinkled leaves have prominent central ribs that may be cut away from the rest of the leaf to be cooked and served like asparagus. The remainder of the leaf is eaten as greens. Swiss chard is unusual in that a single planting can be harvested throughout an entire three-month growing season and even sometimes during a second season as well if flower stalks are removed during the second year. Fordhook Giant is a fine green-leaved variety; Rhubarb is an excellent variety with wine-red leaves. An 8-foot row yields about 7 pounds of leaves during each three-month period.

HOW TO GROW. Swiss chard grows best in soil with a pH of 6.0 to 7.5. In most of the U.S. and southern Canada, where winter frost is expected, sow seeds two to three weeks before the last frost is due; in regions where winter temperatures rarely fall below 25°, sow seeds in fall for harvesting most of the following year. Sow the seeds 1 inch deep and about 4 inches apart in rows 18 to 24 inches apart. When the seedlings are 6 to 8 inches tall, thin them to stand 8 inches apart. The pulled plants may be eaten. Once every four to six weeks scatter 5-10-5 fertilizer around the plants at the rate of 3 ounces per 10 feet of row. The plants are ready to be picked approximately two months after seeds are sown, when outer leaves become 6 to 10 inches tall. Cut off the leaves near the base of the plant with a sharp knife; the undisturbed inner leaves will continue to grow and will be ready for picking a few days later. Pick off and discard any old or tough leaves; if they remain on the plant, they will prevent the plant from producing fresh foliage. Pick off all flower stalks the second year to ensure a second crop.

CHINESE CABBAGE, also called CELERY CABBAGE
Brassica chinensis and *B. pekinensis*

The common name Chinese cabbage is applied to two cool-weather species that are not very closely related to

SWISS CHARD

CHINESE CABBAGE

For climate zones and frost dates, see maps, pages 148-149.

cabbage. Their leaves and stalks have a lettucelike flavor and are eaten raw in salads or cooked. The more common type, *B. pekinensis,* grows 18 to 20 inches tall in a vase-shaped cluster of leaves; when the outer leaves are removed, an oval core of tender, loosely compressed light green leaves is exposed. A good variety is Michihli. A loose-leaved type, *B. chinensis,* grows 12 to 18 inches tall and has edible greenish white stalks; a good variety is Crispy Choy. A 10-foot row yields about 12 heads over three weeks.

HOW TO GROW. Chinese cabbage does best in soil with a pH of 6.0 to 7.5. The soil requires special preparation: dig a 2-inch layer of compost or a 4-inch layer of well-rotted cow manure into a strip 12 to 18 inches wide and 8 inches deep, then add 1 pound of 10-10-10 fertilizer per 10 feet of row. In most of the U.S. and southern Canada, where frost is expected in winter, sow seeds about three months before the first expected fall frost for harvesting in fall. In regions where winter temperatures rarely fall below 25°, sow half the crop in fall to mature in the spring. If Chinese cabbages are sown in spring, they send up inedible seed-stalks as soon as hot weather arrives, even though the plants are young. To sow seeds, group three or four seeds in a spot, setting each group ½ inch deep and 3 to 4 inches apart in rows 18 to 24 inches apart. When the seedlings become 1 inch tall, pull out all but the best plant in each group. When the plants become 6 to 8 inches tall, pull up every other plant, leaving a spacing of 6 to 8 inches. If the soil is very rich, the plants will become overcrowded. When they begin to touch one another, thin again for a final spacing of 12 to 16 inches. The plants that have been pulled in thinning are best taken to the kitchen for cooking. Fertilize every three to four weeks during the growing season, scattering a 12-inch band of 10-10-10 fertilizer along each side of the row at the rate of 5 ounces per 10 feet of row. Both types are ready to be harvested when they are about 15 to 18 inches tall, 80 to 90 days after seeds are sown. To harvest, pull up the plants and cut off the roots; peel away and discard the tough outer leaves.

COLLARD, also called TREE CABBAGE
Brassica oleracea acephala

Collards are generally thought of as vegetables of the Deep South because they are so popular there, but they can also be raised elsewhere. Their alternate name, tree cabbage, is quite descriptive, because when collards are young they do indeed look like cabbages, each collard plant having a rosette of handsome blue-green leaves, but as the plant matures, it does not form a head of tightly compressed leaves as cabbages do; instead, its stems elongate, almost in a treelike fashion, eventually reaching 2 to 4 feet in height. Since each leaf is fully exposed to the sun, it becomes a deep green; it contains more of vitamins A and C than cabbage leaves do, although the taste is similar to but richer than that of cabbage. The leaves are eaten raw in salads or as a cooked vegetable. Collards are also more resistant to heat and cold than cabbages (light frost enhances their flavor). Excellent varieties are Carolina Header, Georgia and Vates. Twenty-five feet of row yields about 20 pounds of collards over a period of three months.

HOW TO GROW. Collards grow best in soil with a pH of 6.0 to 7.5. To reduce the risk of cabbage diseases, seeds should not be sown where other collards, cabbages or any other cabbage relatives have grown within the past three years. In most of the U.S. and southern Canada, where frost is expected in winter, sow seeds in early spring, as soon as the soil can be worked, for harvesting throughout summer and fall. In regions where winter temperatures rarely drop below 25°, sow seeds in late summer or early

COLLARD

fall for harvesting throughout the winter and spring. Group two or three seeds in a spot, setting each group ½ inch deep and 3 inches apart in rows 3 feet apart. When the seedlings become 1 inch tall, cut off all but the strongest plant in each group. When they are 4 to 5 inches tall, pull every other plant. As the plants increase in size, begin to pull every other one until the plants are spaced 3 feet apart (the pulled plants are good to eat). Fertilize every three to four weeks, scattering a 12-inch band of 10-10-10 fertilizer along each side of the row at the rate of 8 ounces per 10 feet of row. Collard roots lie close to the surface of the soil and are easily damaged by cultivation. If it becomes necessary to cultivate, penetrate the soil no deeper than 1 inch. Collards mature fully about three months after seeds are sown, but leaves can be eaten from plants that are only two months old. When picking the leaves, do not disturb the central bud, since the plant continues to send out more and more leaves as the stem grows taller.

CORN
Zea mays

If you can make room for it, sweet corn is the supreme home-grown vegetable simply because growing it yourself is the only way to get corn fresh enough to have full flavor. It loses much of its sweetness within minutes after picking. Most gardeners grow hybrid varieties, which are more vigorous, produce higher yields and have greater disease resistance than standard varieties—but one standard variety, Golden Bantam, is so sweet that it is used as the flavor yardstick by which other sweet corn is measured.

Sweet corn comes with yellow, white, yellow-and-white (called bicolored) or black kernels. The yellow types are most popular; black-kerneled types are rarely offered by seedsmen. In addition to the usual early, midseason and late varieties, there are a number of types known as second-early varieties, which may take about a week longer to mature than early ones. Two excellent standard yellow varieties are Golden Midget, early, and Golden Bantam, midseason. Among the best yellow hybrid varieties are Golden Beauty, Seneca Beauty Hybrid and Seneca Star Hybrid, all early; Early Xtra Sweet Hybrid, Spancross and Sugar and Gold, all second-early; Honey and Cream, Marcross and Tastyvee Hybrid, all midseason; and Burpee's Honeycross, Golden Cross Bantam and Seneca Chief, all late. Two good standard white varieties are Country Gentleman and Stowell's Evergreen, both late; good hybrid white varieties are Silver Sweet, early; Burpee's Snowcrest, midseason; and Silver Queen Hybrid, late. Each cornstalk bears one or two ears of corn. Twenty-five feet of row yields about 40 ears of corn over a period of a week.

HOW TO GROW. Sweet corn grows best in soil with a pH of 6.0 to 7.0. In most of the U.S. and southern Canada, where frost is expected in winter, sow seeds outdoors in spring when all danger of frost is past. Oldtimers used to say, "Plant corn when the leaves on the oak trees are as big as a mouse's ear." This advice is still accurate, since oak leaves generally become that big soon after the last spring frost. If the spring is cold and damp, however, the seeds may rot in the ground, so most gardeners make a second planting about two weeks after the first. These two plantings often mature about the same time because corn grows slowly in cold weather. Plant second-early, midseason and late varieties at the same time as early varieties, or make successive plantings of early varieties every two weeks until three months before the first fall frost is due. Either method assures a continuing supply of fresh ears throughout summer and early fall. In frost-free regions sow seeds any time average temperatures are expected to stay

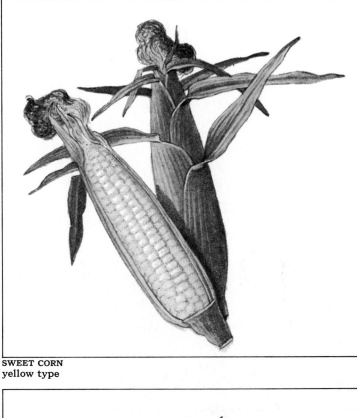

SWEET CORN
yellow type

SWEET CORN
bicolored type

For climate zones and frost dates, see maps, pages 148-149.

within a range of 40° at night to a high of 85° during the day over a period of two months.

Corn is planted either in clumps, or hills, or in rows. Because corn is pollinated by wind-borne pollen, which cannot travel far, sow the seeds in four rows or four hills to make it easier for the plants to pollinate one another during the blossoming season. In spring, when the soil is moist and cool, sow the seeds about 1 inch deep. During the summer, when there is less moisture near the surface of the soil, sow the seeds about 2 inches deep. When planting corn in rows, sow the seeds 3 to 4 inches apart. Space rows of early corn 2 feet apart; space rows of the taller-growing midseason and late varieties about 3 feet apart. When the plants are 2 to 4 inches tall, cut off all but the strongest ones for a final spacing of about 12 inches. When planting hills of corn, group six seeds in a spot, setting each group 2 feet apart in both directions. When the plants are 2 to 4 inches tall, cut off all but the three strongest plants in each hill. Some gardeners follow the Indian custom of planting a few pumpkins *(page 104)* and winter squashes *(page 106)* 10 feet apart among their corn.

Fertilize the plants when they are 2 to 4 inches tall and again when they are 8 to 10 inches tall, scattering a 6-inch band of 10-10-10 fertilizer at the rate of 5 to 8 ounces of fertilizer for every 10 feet of row. Water the plants whenever they show any signs of wilting, and keep the soil moist when the tassels appear because this is the time when the ears are forming. Pull weeds when they become 2 inches tall; otherwise, cultivate the soil no deeper than 1 inch—corn roots are very shallow and are easily injured.

Corn is ready for picking when the silks at the ends of the ears turn brown and the ears themselves are full and firm. The kernels should be juicy and plump, but not overgrown, or the flavor will be impaired. Early varieties mature in about 65 days; second-early varieties in about 70 days; midseason varieties in about 75 days; and late varieties in 85 to 95 days. Corn planted in midsummer usually takes about two weeks longer to mature than corn planted in spring. To pick corn, pull the ear down and twist it free.

Unused corn seeds will keep for about two years.

CUCUMBER
Cucumis sativus

Cucumbers grow on vines that spread 6 to 8 feet if allowed to trail on the ground. In small gardens they can be trained to climb a fence and often produce better-shaped fruit than trailing plants. Cucumbers have dark green leaves 4 to 6 inches across, which have to be rolled to one side to reveal the cucumbers themselves, and bear yellow flowers about 1 inch across. A cucumber grows behind each female flower. To increase the production of female flowers—and, therefore, of cucumbers—plant breeders offer seed packages of all-female varieties that include a few green seeds instead of the normal beige. The green seeds produce cucumbers with many male flowers, and one of these seeds is planted with every five or six female seeds; the combination ensures pollination of a large number of female flowers—and tremendous crops.

There are two kinds of cucumbers for home gardens, slicing cucumbers and pickling cucumbers. Slicing cucumbers may also be used for pickling, but are best eaten raw. Recommended slicing varieties are Burpee Hybrid, Burpless Hybrid, Marketmore, Poinsett, Spartan Valor Hybrid, and Victory Hybrid; the last two are all-female. Good pickling varieties are Bravo Hybrid (all-female), Ohio **MR** 17 and Wisconsin SMR 18. A 10-foot row yields about 30 pounds over a period of six weeks.

HOW TO GROW. Cucumbers grow best in soil with a pH

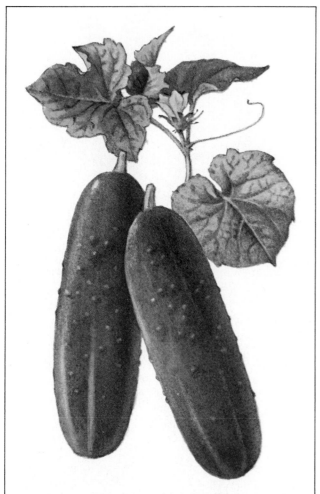

CUCUMBER

of 5.5 to 7.0. In most of the U.S. and southern Canada, where frost is expected in winter, sow seeds indoors in peat pots about three weeks before night temperatures can be depended upon to stay above 55°. Set two seeds ½ inch deep in each pot. When the seedlings are 1½ inches tall, cut off the weaker of the two. After all danger of frost is past, set the seedlings, pots and all, into the garden. Seeds may also be sown directly outdoors when all danger of frost is past. In frost-free areas, sow seeds outdoors in fall for harvesting in winter and spring.

Cucumbers are usually grown in clumps, or hills. To prepare a hill, dig a hole 1 foot deep and 2 feet across. Dig into the bottom of each hole 2 to 4 inches of compost or well-rotted cow manure, then return the removed soil, forming a mound about 4 inches high. For trailing plants, space hills 4 feet apart; for climbing plants, space individual plants 10 to 12 inches apart. Set plants started indoors two or three to a hill in a 1-foot circle on top of each mound. When sowing seeds directly outdoors, plant five or six seeds ½ inch deep, also in a 1-foot circle, on top of each mound. When the seedlings become 2 to 3 inches tall, cut off all but the three strongest. If an all-female variety is planted, mark where the male seeds are sown to avoid removing all of the male plants when thinning.

To protect young plants from hard rain, insects and late frosts, and to provide warmth needed to make them grow faster, cover them with translucent wax-paper cups available from garden centers. For climbing plants, hang plastic garden netting on metal fence posts (drawing, page 30).

Scatter a handful (about ⅓ cup) of 5-10-5 fertilizer around each hill every two to three weeks, and water enough to keep the soil moist and prevent wilting. Cucumber roots lie close to the surface of the soil and are easily damaged. If it becomes necessary to cultivate, penetrate the soil no deeper than 1 inch. Cucumbers are usually ready to be harvested about two months after seeds are sown—slicing cucumbers when they are 6 to 8 inches long, pickling cucumbers when they are 1½ to 3 inches long. Cut all cucumbers while they are dark green even if you cannot use them—if the fruit turn yellow, the plants stop producing. Cut the cucumbers from the vines; if you break them off, the vines may become entangled and break.

Unused cucumber seeds will keep for about five years.

E

EGGPLANT
Solanum melongena esculentum

Eggplants, widely grown as a meat substitute in the Mediterranean region and India, are less common here because they require very warm weather—night temperatures of at least 55° and day temperatures of about 80° or more—for a period of about two and a half months after young plants are set into the garden.

The plants usually grow 2 to 3 feet tall (in the tropics they become as much as 8 feet tall), with each plant bearing four or more fruit. The fruit may be up to 10 inches long when mature, but are far more flavorful when eaten at the half-grown stage. Good varieties are Black Beauty, Black Magic Hybrid, Burpee Hybrid and Early Beauty Hybrid; a variety particularly suited to the Southwest and South is Florida Market. A 6-foot row yields about 12 to 30 fruit over a period of six weeks.

HOW TO GROW. Eggplants grow best in soil with a pH of 5.5 to 6.5. To reduce the risk of damage from pests and diseases carried by other plants, eggplants should not be located where tomatoes, potatoes or other eggplants have grown within the past three years. Start seeds indoors or in a hotbed 10 weeks before night temperatures can be de-

EGGPLANT

For climate zones and frost dates, see maps, pages 148-149.

pended upon to stay above about 55°; sow the seeds ½ inch deep. When the seedlings become 1 to 1½ inches tall, transplant them to 3- or 4-inch peat pots. Keep the plants warm and moist at all times so that growth will continue unchecked; stunted plants do not bear fruit. When night temperatures can be depended upon to stay above 55°, set the plants into the garden. Space them 2½ to 3 feet apart in rows 2½ feet apart. Pour around each plant 1 pint of water-soluble fertilizer diluted to half the strength recommended on the label. Protect the plants from cutworms by slipping over them paper cups that have had the bottoms removed. Whenever there has been less than 1 inch of rain in a week, give the plants a thorough soaking. Every three to four weeks scatter ¼ cup of 5-10-5 fertilizer in a 1½-foot band around each plant. The fruit is ready to pick when it is about 6 inches long and very shiny —about 145 days after seeds are sown or about 70 days after seedlings are transplanted to the garden. If the fruit gets dull, it is overripe. Pick all fruit before they mature; otherwise the plants stop producing. The stems of the fruit are tough and should be cut with shears.

ENDIVE

Endive, also called curly endive, and escarole (*Cichorium endiva*)

Endive and escarole look like lettuce, forming a nearly flat rosette of leaves up to 18 inches across when mature, but their flavor is somewhat sharper. (Belgian endive is actually witloof, a plant very different in appearance.) Endive has slender leaves with wavy edges, escarole rather broad flat leaves. Both types must be blanched—that is, shielded from light—to remove some of the green from the leaves, making them yellowish green, and also to eliminate a bitter flavor. Good varieties of endive are Green Curled and Salad King; good escarole varieties are Florida Deep Hearted and Full Heart Batavian, also called Broad-Leaved Batavian. A 15-foot row yields about 15 heads over a period of two weeks.

HOW TO GROW. Endive and escarole grow best in soil with a pH of 5.8 to 7.0. In most of the U.S. and Canada, where frosts are expected in winter, sow seeds in midsummer about three months before the first fall frost is due; the crop will be ready in fall. Because the plants tolerate frosts, seeds may also be sown in very early spring for summer harvesting, but the plants will not be as satisfactory as those grown for harvesting in fall. In regions where winter temperatures rarely fall below 25°, sow seeds in fall for harvesting during winter and spring. To sow seeds, group three or four in a spot, setting each group ¼ inch deep and 8 to 12 inches apart in rows 2 feet apart. When the seedlings become 1 to 2 inches tall, pull out all but the strongest plant in each group. If you have limited space in your garden, you can sow the seeds in a seedbed; then when the seedlings are 2 to 3 inches high, move them to the garden where other plants, such as beans, have already been harvested. The roots lie close to the surface of the soil and are easily injured by cultivation. If it becomes necessary to cultivate the plants, penetrate the soil no deeper than an inch. Fertilize every three to four weeks, scattering a 6-inch band of 10-10-10 fertilizer along each side of the row at the rate of 5 ounces per 10 feet of row. When the plants are well formed and about 15 inches across, blanch the plants for two to three weeks. To blanch, gather the long outer leaves together over the crown of each plant, holding the leaves in place with a thin rubber band. Or cover the plants with a wide board, laying it directly over the tops of the plants; the foliage is nearly flat on the ground anyway and will not be damaged. Start the

ENDIVE

blanching process when the leaves are dry. To prevent decay, uncover the plants and let them dry out after any rainfall. Endive and escarole are ready for harvesting about three months after seeds are sown. They can be picked and eaten at any size after they have been blanched. Unused seeds will keep for about five years.

ESCAROLE See Endive

G

GUMBO See Okra

K

KNOB CELERY See Celery

KOHLRABI
Brassica oleracea caulorapa

Kohlrabi is a cool-weather relative of cabbage that is grown for the swollen turnip-shaped portion of its stems, which rests on the surface of the ground. The stems have creamy white interiors and are eaten raw in salads or cooked as a vegetable. Good varieties are Early White Vienna, which has greenish white skin, and Early Purple Vienna, which has a light purplish skin. A 12-foot row yields about 24 kohlrabi stems over a period of four weeks.

HOW TO GROW. Kohlrabi grows best in soil with a pH of 6.0 to 7.5. To reduce the risk of cabbage diseases, seeds should not be planted where other kohlrabi plants, cabbages or any other cabbage relatives have grown within the past three years. In most of the U.S. and southern Canada, where frost is expected in winter, sow seeds in early spring as soon as the soil can be worked and continue to sow at two-week intervals until one month before the first fall frost is expected. In regions where winter temperatures rarely fall below 30°, sow seeds at two-week intervals beginning in late summer for harvesting in winter and spring. Group three or four seeds in a spot, setting each group ½ inch deep and 6 inches apart in rows 18 to 24 inches apart. When the seedlings become 1 to 1½ inches tall, cut off all but the strongest in the group. At three-week intervals scatter 10-10-10 fertilizer around the plants at the rate of 5 ounces per 10 feet of row. Keep the soil constantly moist. Kohlrabi roots lie close to the surface of the soil and are easily damaged by cultivation. If it becomes necessary to cultivate, penetrate the soil no deeper than 1 inch. Kohlrabi is ready to be picked about eight weeks after sowing, when the stems are about the size of an apple 2 to 2½ inches across. If the stems are allowed to grow older, they become tough and stringy. Harvest the plants by cutting the stems at soil level so as not to disturb the roots of adjacent plants.

L

LEEK
Allium porrum

Leeks are mild-flavored relatives of the onion, but unlike onions they do not form bulbs. Instead, they develop edible 6- to 10-inch-long cylindrical stems, as much as 2 inches in diameter, that are eaten raw or cooked. The plants grow 2 to 3 feet tall and bear flat straplike, rather tough leaves that are sometimes used in cooking. Young leeks share the name scallions with bunching onions *(page 99)*. The most widely grown variety of leeks is Broad London, sometimes sold as Large American Flag. A 10-foot row will yield about 30 leeks over a period of four weeks.

HOW TO GROW. Leeks grow best in soil with a pH of 6.0 to 8.0. In regions where winter frosts are expected, sow seeds indoors about two months before the last frost is

KOHLRABI

For climate zones and frost dates, see maps, pages 148-149.

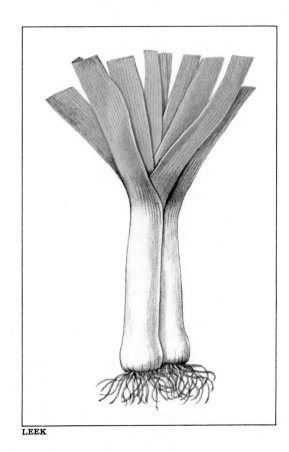

LEEK

LETTUCE
Top to bottom: head, loose-leaf and butterhead types

due; the plants will be ready to harvest in midsummer. When the seedlings are 4 to 5 inches tall, dig a trench in the garden, making it 6 inches deep and 6 inches across. Set the seedlings at the bottom, spacing them 4 to 6 inches apart. If more than one row is planted, space the trenches 15 to 18 inches apart from center to center. Seeds can also be sown directly in the garden when night temperatures can be depended upon to stay above freezing, but the plants will mature later than those sown indoors. In frost-free regions, sow seeds in the garden in late summer when maximum daytime temperatures average no higher than 80°. Prepare a trench as described above for transplanting seedlings. Group three or four seeds in a spot at the bottom of the trench, spacing the groups 4 to 6 inches apart; cover the seeds with ⅛ inch of soil.

The trench that leeks are grown in—whether from seeds or seedlings—is not filled all at once but gradually over the season. When the plants are 3 to 4 inches tall, push soil around the stems up to the leaves. As the plants grow, continue to surround the stems with soil until the plants reach maturity and the trench becomes filled in. The soil will blanch the stems—that is, turn them white by withholding light—and will also make them tender. Fertilize every three to four weeks, scattering a 6- to 8-inch band of 5-10-5 fertilizer along each side of the row at the rate of 5 ounces of fertilizer per 10 feet of row. Leeks mature in about four and one half months but are edible at any earlier stage. Unused leek seeds will keep for about two years.

LETTUCE
Common lettuce *(Lactuca sativa);* romaine lettuce, also called Cos lettuce *(L. sativa longifolia)*

Lettuce exists in many types. Common lettuce comes in three forms—head lettuce, loose-leaf lettuce and butterhead lettuce. Head lettuce has green outer leaves that spread 12 to 15 inches across from a tightly compressed pale green central ball. It is widely sold in supermarkets but is easily grown in home gardens only in limited areas, such as around the Great Lakes in summer and in the Imperial Valley of California in winter, where temperatures stay within a range of 35° at night to 80° in the daytime for at least two months. Recommended varieties are Iceberg, Great Lakes, Imperial 44, Ithaca and Oswego. A 15-foot row yields about 15 heads over a period of two weeks.

All other types of lettuce listed here grow well anywhere in the U.S. and southern Canada. Loose-leaf lettuce forms a rosette of tender green leaves 8 to 12 inches across. Recommended varieties are Oakleaf, Black-Seeded Simpson, Prizehead, Ruby (its leaves are tinged with red) and Salad Bowl; a 15-foot row yields about 2½ pounds over a period of three weeks. Butterhead lettuce has a softly compressed head, also 8 to 12 inches across, consisting of green outer leaves and pale green to yellow inner leaves. Recommended varieties are Buttercrunch, Dark Green Boston, Deer Tongue (also called Matchless), Fordhook Summer Bibb and Tender Crisp; a 15-foot row yields about 15 heads over a period of three weeks.

Romaine lettuce forms a vase-shaped and tightly compressed head of foliage about 10 inches tall. Its leaves have a more piquant flavor than those of other lettuces. Recommended varieties are Parris Island Cos, Dark Green Cos and Paris White. A 15-foot row yields about 20 heads over a period of three weeks.

HOW TO GROW. Lettuce should be planted where it gets full sun in spring and fall but partial shade during the hottest part of summer. It grows best in soil with a pH of 6.0 to 7.0. In most of the U.S. and southern Canada, where frosts are expected in winter, sow seeds indoors or in a hotbed about a month before the last frost is expected. When the seedlings become 2 to 3 inches tall and night temperatures are no longer likely to fall below 25°, move the plants to the garden, spacing head lettuce about 12 inches apart, other types 6 to 8 inches apart; the rows should be 1½ feet apart. For additional crops of lettuce, sow seeds directly in the garden in early spring as soon as the soil can be worked, and continue to sow every two weeks until about two months before maximum daytime temperatures are expected to average about 80°; where maximum summer temperatures do not average 80°, continue successive plantings until two months before the first frost is expected in fall. In regions where winter temperatures rarely fall below 25°, start successive plantings in early fall when maximum daytime temperatures average below 80°. Such plantings will be harvested in late fall, winter and spring; make the final planting two months before maximum daytime temperatures are expected to average above 80°.

In all areas, sow seeds sparingly, ½ inch deep, and thin the plants to the spacing given for transplanted seedlings. The pulled plants are good to eat. Fertilize every three weeks, scattering a 6- to 8-inch band of 10-10-10 fertilizer along each side of the row at the rate of 5 ounces to every 10 feet of row. Keep the soil moist, but do not wet the foliage any more than necessary. Head lettuce matures 10 to 11 weeks after seeds are sown, loose-leaf lettuce 6 to 7 weeks, butterhead lettuce 9 to 10 weeks and romaine lettuce 11 to 12 weeks. However, the leaves of all types are edible at any stage.

Unused lettuce seeds will keep for about six years.

M

MELON
Cantaloupe, also called muskmelon *(Cucumis melo reticulatus);* casaba, Crenshaw, Persian and honeydew melons *(C. melo inodorus);* watermelon *(Citrullus vulgaris)*

Though melons are thought of as fruits, they are treated like vegetables in the garden. They grow on vines that creep along the ground for 6 to 10 feet or more and are usually planted in clumps on mounds, or hills, of soil.

Melons can be grown successfully in regions where minimum night temperatures average no lower than 55° and minimum daytime temperatures are no lower than 80° throughout the growing seasons. The length of the growing seasons, from the time the seeds are sown until the plants are harvested, is listed for each recommended variety. Because long growing seasons are required for most casaba, Crenshaw, Persian and honeydew melons, all but a few of the varieties that are listed in these categories are generally grown only in the southwestern, south-central and southern parts of the U.S.

Most cantaloupes have orange flesh, but some have lime-green flesh even when ripe. Some excellent varieties are Burpee's Fordhook Gem, green flesh, and Burpee Hybrid, orange flesh, both 82 days; Delicious 51, orange flesh, 86 days; and Mainerock Hybrid, orange flesh, 75 days. Two hills of cantaloupe yield a harvest of about eight fruit over a period of three weeks.

Casaba, Crenshaw, Persian and honeydew melons are closely related to cantaloupes, but generally ripen later. Excellent varieties for warm regions are Golden Beauty Casaba, white flesh, 120 days; Honey Dew, white flesh, 110 days; and Persian, orange flesh, 120 days. Three good varieties that succeed in shorter growing seasons are Burpee's Early Hybrid Crenshaw, pink flesh, 90 days; Honey Mist, green flesh, 92 days; and Sungold Casaba, white flesh, 85 days. Two hills planted with these types yield six fruit over a period of four weeks.

ROMAINE LETTUCE

For climate zones and frost dates, see maps, pages 148-149.

Watermelons are available as small round "icebox-sized" melons weighing from 4 to 6 pounds, large oblong types weighing 20 to 40 pounds and seedless watermelons weighing 10 to 20 pounds (seedless watermelons sometimes have a few white seeds). Excellent small watermelons are Sugar Baby and the slightly heavier, sweeter yellow-fleshed Yellow Baby from China, both 75 days; large watermelons, Charleston Gray, 85 days, and Crimson Sweet, 80 days; seedless watermelons, Burpee Hybrid Seedless and Triple Sweet Seedless, 80 days. Two hills yield six large or 12 small watermelons over a period of three weeks.

HOW TO GROW. All melons grow best in light, sandy soil with a pH of 6.0 to 7.5. In regions where minimum night temperatures average above 55° for less than three months, sow seeds of all types of melons indoors or in a hotbed about one month before night temperatures can be depended upon to stay above 55° and daytime temperatures above 80°, then set them into the garden when the required temperatures are reached. Elsewhere, sow seeds of all types directly outdoors when the temperatures reach 55° at night and 80° in the daytime. When growing seedless varieties, always plant a normal seed-type watermelon nearby to pollinate them so they can produce fruit; set the seed types in a separate hill to make sure that all of the pollinating plants are not pulled up when thinning. The variety Sugar Baby is often used for this purpose because it produces an abundance of pollen.

To prepare a hill for melons, dig a hole about 1 foot deep and 2 feet across; dig into the bottom of the hole a 4- to 6-inch layer of compost or well-rotted cow manure. Replace the topsoil until it forms a gentle mound about 4 inches high. Space hills for large watermelons about 10 feet apart, for all other melons 4 to 6 feet apart.

Transplant seedlings started indoors two to a hill. When sowing seeds directly outdoors, plant six to eight seeds on top of each hill in a circle about 12 inches across; set the seeds about ½ inch deep. When the seedlings appear, cut off all but the two best. To protect seedlings from hard rain, insects and late frosts, and provide warmth to speed growth, cover them with translucent wax-paper caps available for that purpose from garden supply stores. Fertilize every two weeks, scattering about ⅓ cup of 5-10-5 fertilizer around each hill. Water the plants in dry weather. Because melons lie on the ground, a mulch of old hay or straw helps prevent rot. Also, melon roots are shallow and are easily damaged by cultivation; if a mulch is not used, hoe no deeper than 1 inch when weeding. Do not move the vines; they too are easily injured.

Cantaloupes should be picked at what is called the "slip" stage, when a slight pressure at the point where the stem joins the melon causes the melon to slip off the vine. All other melons are still firmly attached to their vines at harvesttime, so other yardsticks must be used: casaba and honeydew melons are ripe when the skin turns yellow; Crenshaw and Persian melons when they develop a fruity scent; watermelons when a rap on the fruit creates a dull rather than a sharp sound.

Fruits that start to grow after midsummer will not have time to mature and should be removed; this thinning will direct nourishment toward fruits that are developing.

Unused melon seeds keep for about five years.

O

OKRA, also called GUMBO
Hibiscus esculentus

Okra is a warm-weather plant grown for its immature seed pods, which are delicious fried or cooked in soups and stews. The plants grow 3 to 5 feet tall and bear maroon-

CANTALOUPE

centered pale yellow flowers about 2 inches across. The flowers quickly develop into slender pointed seed pods that become 7 to 9 inches long when fully mature, but are most flavorful and tender if picked when only 2 to 3 inches long. Good varieties are Clemson Spineless, Dwarf Green Long Pod, Emerald and Louisiana Green Velvet. An 8-foot row yields about 5 pounds of pods every two weeks until frost if the pods are picked daily.

HOW TO GROW. Okra grows best in soil with a pH of 6.0 to 8.0. In most of the U.S. and southern Canada, where frost is expected in winter, sow seeds indoors or in a hot-bed about a month before night temperatures are expected to stay above 50°. Sow the seeds in peat pots, two seeds per pot. When the seedlings are 1 inch tall, cut off the weaker one in each pot. When night temperatures no longer fall below 50°, set the plants, pots and all, into the garden, spacing them 18 inches apart in rows 3 feet apart. In frost-free regions, sow seeds directly in the garden when night temperatures are expected to stay above 50°. Group three or four seeds in a spot, setting each group ½ inch deep and 18 inches apart in rows 3 feet apart. When the seedlings become 1 inch tall, cut off all but the strongest in each group. Fertilize twice—when the plants are 8 to 12 inches tall and again just as they begin to blossom; scatter a 12-inch band of 5-10-5 fertilizer around each plant at the rate of 5 ounces to every 10 feet of row. Okra begins to produce pods about 60 days after seeds are sown. The pods develop very rapidly and should be picked daily within a few days after the flower petals have fallen, whether the pods are to be used or not. If pods are allowed to ripen, the plants cease to produce.

ONION
Allium cepa

The two most popular types of onions are ordinary onions and bunching onions. Ordinary onions are bulbs that lie on or close to the surface of the soil and bear 18-inch-high hollow rounded blue-green leaves. They come with white, yellow or red skins and include Spanish, Bermuda and Portugal onions, which are mild-flavored large bulbs that generally do not keep as long as other ordinary onions. (The stronger the flavor of the onion the longer it keeps.) Excellent white varieties are Crystal White Wax, an early Bermuda onion, which keeps only about a month; and White Portugal, also called Silverskin, and White Sweet Spanish, both of which keep about two months. Typical yellow varieties are Downing Yellow Globe, Early Yellow Globe and Ebenezer, which keep three months or more; and Utah Sweet Spanish, a large variety that keeps about two months. Among the good red-skinned varieties are Red Burgundy, which keeps about two months; and Red Weathersfield and Southport Red Globe, which keep three months or more. A 10-foot row of onions yields about 10 pounds over a period of 10 weeks.

Bunching onions, also called green onions or scallions, form mild-flavored thick stems instead of bulbs. The plants are cold resistant and can be grown over winter to provide tender young stems in early spring. Bunching onions keep only one to two weeks. Good varieties are Evergreen White Bunching and Hardy White Bunching. A 10-foot row yields about 10 bunches over four weeks.

HOW TO GROW. Onions grow best in soil with a pH of 5.5 to 7.0. They can be started from seeds, small bulbs (called sets) or small plants. Because onions grown from seeds need five months to mature and because the plants are tedious to weed when small, it is easier and faster to use sets or young plants. In most of the U.S. and southern Canada, where frost is expected in winter, plant onions in spring as

WATERMELON
midget type

OKRA

For climate zones and frost dates, see maps, pages 148-149.

soon as the soil can be worked for harvesting in summer and fall (or the following spring for bunching onions). In frost-free regions, plant in fall for harvesting in spring.

Plant sets or young plants about 2 inches deep and 2 to 4 inches apart; if 2-inch spacing is used, pull and serve alternate plants when they are 6 inches or more tall. If seeds are used, sow them ½ inch deep in rows 12 to 18 inches apart. When the seedlings are 2 inches tall, thin them to stand 2 inches apart. When they become about 6 inches tall, pull up every other plant (they are edible), making the final spacing 4 inches apart.

Fertilize onion plants twice—when they are about 6 inches tall and again when they are about 12 inches tall; scatter a 4-inch band of 5-10-5 fertilizer along each side of the row at the rate of 3 ounces to 10 feet of row. Onions have shallow roots and need constant moisture.

Ordinary onions can be harvested about five months after the seeds are sown or about three and one half months after sets or young plants are planted. When the leaves begin to turn yellow, bend the stems into a nearly horizontal position; this stops growth and allows the bulbs to ripen. Pull away any mulch and part of the soil from around the bulbs until they are half exposed. When the leaves turn brown, lift the bulbs from the soil. Cut the tops off 1 inch from the bulbs and spread the bulbs out to dry for a week or more. Or braid the tops of small bulbs and hang them to dry (bulbs over 2 inches in diameter are usually too heavy to cling securely in a braid). Bunching onions should be harvested as needed because they keep for such a short period.

Use fresh onion seeds each year.

P

PARSNIP
Pastinaca sativa

Parsnips are grown for their delicate-tasting roots, which grow up to 15 inches long and 3 to 4 inches across at the top. The plants are biennials but are grown as annuals and should be harvested before the second year's leaves form. Never pick any wild plant that looks like parsnip; the poisonous water hemlock is easily mistaken for it. Good varieties of parsnips are All American, Guernsey, Model and Hollow Crown. A 15-foot row yields about 15 pounds in a single harvesting.

HOW TO GROW. Parsnips grow best in a soil of pH 5.5 to 7.0 that has been cultivated about 18 inches deep and freed of all rocks to allow the roots to grow smooth and straight. In most of the U.S. and southern Canada, where frost is expected in winter, sow seeds as soon as the ground can be worked in early spring. In regions where winter temperatures rarely fall below 25°, sow seeds in early fall for harvesting the following spring.

Parsnips have a hard time pushing their way up through the ground, so sow the seeds thickly to allow for those that fail to germinate. Set them ½ inch deep in rows 2 to 2½ feet apart, including in each row a few seeds of radish (*page 104*)—radishes grow quickly and mark the row until the parsnips sprout. To keep the soil soft and moist, and help the seeds germinate, cover them with vermiculite, sifted compost or a light layer of grass clippings. Firm the covering well and water. When the plants are 1 inch tall, thin them to stand 3 to 6 inches apart. Fertilize once a month during the growing season, scattering a 6- to 8-inch band of 5-10-5 fertilizer along each side of the row at the rate of 5 ounces to every 10 feet of row. The plants mature about four months after seeds are sown. Parsnips planted in spring may be harvested in early fall and stored in the refrigerator until needed, but their flavor is enhanced

ONION
Top left and right: bulbous types   *Bottom:* bunching type

if the roots are left in the ground all winter. If you want to dig them up from time to time in winter, cover the plants with hay or other mulch to keep the ground from freezing and make it easy to get to the roots. No mulch is needed if you harvest the whole crop the following spring.

Use fresh parsnip seeds each year.

PEA
Green pea, also called garden pea or English pea *(Pisum sativum);* edible-pod pea, also called Chinese snow pea or sugar pea *(P. sativum macrocarpum)*

Peas are among the earliest vegetables to be picked each year and should be eaten almost immediately because, like corn, they lose their sweet flavor very rapidly. All grow on vines and do best in cool weather.

Green peas are classified as smooth or wrinkled, according to the way their seeds look when dried. Since wrinkled varieties are sweeter than smooth ones, only wrinkled varieties are recommended here. The length of the growing season for each variety, from the time seeds are sown until the plants are ready to be harvested, follows the varietal name. Fine tall-growing types are the 4- to 5-foot Alderman, also called Tall Telephone, 74 days; and three 2½-foot varieties—Freezonian, 62 days; Lincoln, 67 days; and Wando, 65 days. Dwarf types are the 15-inch Little Marvel, 64 days; and the 18-inch Sparkle, 59 days. A 15-foot row of green peas yields about 5 pounds a week.

Two fine varieties of edible-pod peas are the 2- to 2½-foot Dwarf Gray Sugar, 65 days; and the 4½-foot Mammoth Melting Sugar, 74 days. A 15-foot row of edible-pod peas yields about 3 pounds over a period of one week.

HOW TO GROW. Peas grow best in soil with a pH of 6.0 to 7.5. To prepare the soil for a double row with 8 inches of space down the middle, dig a flat-bottomed trench about 10 inches wide and 2 inches deep. If additional double rows are needed, space the trenches 3 feet apart. Dust the bottom of each trench with a low-nitrogen fertilizer such as 5-10-10 at the rate of 2 ounces for every 10 feet of row, and rake it into the bottom of the trench.

Low-nitrogen fertilizer is used with peas because they, like other legumes, draw nitrogen from the atmosphere with the help of soil bacteria. To make the bacteria available, dust seeds before planting with commercial bacteria cultures available at garden supply centers.

In most of the U.S. and southern Canada, where frost is expected in winter, sow seeds as soon as the soil can be worked in spring, and continue to sow every 10 days until 60 days before maximum daytime temperatures are expected to average 75°. If temperatures rise above 75° before the last crop matures, the plants may fail; an exception is the sturdy Wando variety of green pea, which withstands temperatures as high as 80°. In regions where winter temperatures rarely fall below 30°, start successive plantings in fall and continue until 60 days before maximum daytime temperatures are expected to average 75°.

Plastic garden netting hung on metal posts makes a good support for the plants *(drawing, page 30).* When the vines become 15 inches high, run twine from post to post outside the vines on each side of the fence at a level of about 12 inches from the ground. Repeat this procedure 24 inches from the ground and later at the top of the fence. Dwarf varieties do not need fences; instead, pile up 3 or 4 inches of soil on each side of the row. To keep away birds, cover the plants with plastic netting.

Pods of green peas should be picked while they are firm but still succulent, before they become yellowish or shriveled. Edible-pod peas should be picked while the pods are still flat and the peas within are barely discernible.

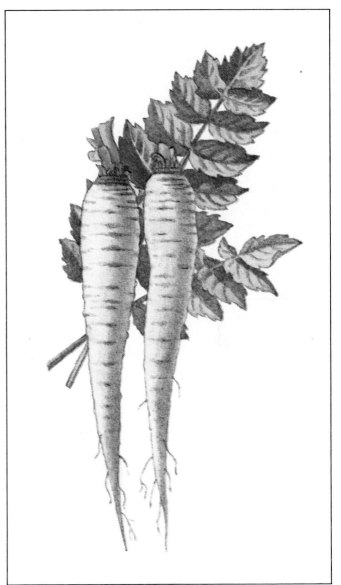

PARSNIP

For climate zones and frost dates, see maps, pages 148-149.

101

Since pea pods are firmly attached to their vines, hold the vine with one hand and pull the pod with the other.

Unused pea seeds will keep for about two years.

PEPPER
Sweet pepper, also called bell pepper and green pepper, and hot pepper *(Capsicum frutescens)*

Both sweet peppers and hot peppers are warm-weather perennial shrubs from the tropics, but are treated as annuals in gardens. The plants usually grow about 2 feet tall with an equal spread.

The fruit of sweet peppers grow 3 to 4 inches long and 2½ to 3 inches wide; they are often harvested while still green and crisp and are eaten either raw or cooked; if allowed to ripen, they turn red or yellow and may become slightly soft, but the flavor is unchanged. Excellent varieties are Ace Hybrid, Bell Boy Hybrid and Yolo Wonder, which turn red when ripe; and Golden Calwonder, which becomes a rich, golden yellow.

Hot peppers vary greatly in size and shape. Some are almost cherrylike, others are up to a foot long and tapering. They are green when they first appear, but quickly turn red or yellow. All have a pungent flavor and are eaten fresh, cooked or pickled. Some varieties, such as cayenne, can be dried and ground. Good hot peppers are Long Red Cayenne, Large Cherry and Tabasco, all red varieties; and Hungarian Yellow Wax, a yellow variety.

A 10-foot row of peppers yields about 6 pounds over a period of six weeks.

HOW TO GROW. Peppers grow best in soil with a pH of 5.5 to 7.0. In most of the U.S. and southern Canada, where winter frosts are expected, start seeds indoors in spring six to eight weeks before minimum night temperatures are expected to average above 55°. Sow the seeds in a flat and transplant the seedlings to individual pots when they are about 1 inch tall. Or sow two or three seeds in individual pots, and when the seedlings become an inch tall, cut off all but the strongest one in each pot. The plants need indoor temperatures of 70° to 80°. Wait until at least two weeks after outdoor temperatures can be relied upon to stay above 55° before transplanting the seedlings to the garden; then set them 18 to 24 inches apart in rows 2 to 3 feet apart. In frost-free areas, sow seeds directly outdoors in spring or fall. To prevent cutworm injury, cover the plants with paper cups that have had the bottoms removed. Fertilize twice—when the plants are about 8 inches tall and again when they are about 12 inches tall—scattering an 8- to 12-inch band of 5-10-5 fertilizer along each side of the row at the rate of 5 ounces to 10 feet of row. Both sweet peppers and hot peppers are edible and flavorful at all stages of their growth. Ace Hybrid requires about 60 days to mature to its full size after it is set out in the garden; Bell Boy Hybrid, 70 days; Yolo, 72 days; Golden Calwonder, 75 days; Long Red Cayenne, 70 days; Large Cherry, 69 days; Tabasco, 92 days; and Hungarian Yellow Wax, 70 days. When picking peppers, cut them from the plant instead of pulling them; the branches are extremely brittle and will break easily if pulled.

If blossoms and young fruits form during a period of low humidity, they will fall to the ground, but if the growing season is long enough and the level of humidity increases, the plants will flower and fruit again. Unused pepper seeds will keep for about two years.

POTATO
Solanum tuberosum

Potatoes are not a crop for most home gardeners to grow in large quantities—but young potatoes, an inch or

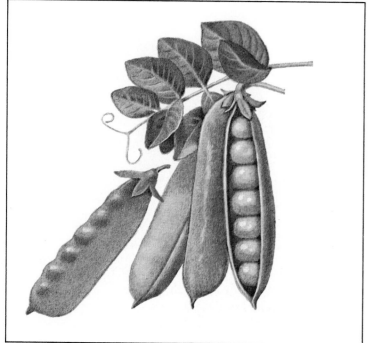

GREEN PEA

two across, have a special flavor and are worth growing. Most varieties have brown skins, but some have reddish or bluish skins. The flesh of potatoes is usually white or cream-colored. The variety to plant depends upon where you live and whether you want early potatoes for eating in summer and fall or late potatoes for storage. In most of the U.S. and southern Canada, where frost is expected in winter, early varieties are Early Gem, Irish Cobbler and Norgold Russet; late varieties are Katahdin, Kennebec and the most famous potato for baking, Russet Burbank, sometimes called Idaho or Idaho Baker. In frost-free regions only one crop is grown each year; good varieties are Irish Cobbler and Russet Burbank. Twenty-five feet of row yields 25 to 50 pounds of potatoes at one harvesting.

HOW TO GROW. Potatoes grow best in light, sandy soil with a pH of 4.8 to 6.5. In most of the U.S. and southern Canada, where frost is expected in winter, plant potatoes in early spring as soon as the soil can be worked. In parts of the South and Southwest where some frost is expected in winter, plant potatoes in January or February so that they can be ready for harvest in spring before daytime temperatures are expected to average 90° or more. In frost-free regions, plant potatoes any time from early fall until midwinter for harvesting in winter and spring.

Potatoes are grown from pieces of seed potatoes, which are specially developed for planting; be sure to buy those labeled "certified"—they are relatively free of diseases and often produce double the crop obtainable with noncertified potatoes. Do not plant potatoes bought at a food store —they may have been shipped in from another part of the country and may not grow well where you live. They may also have been treated with a chemical that inhibits sprouting and will prevent them from growing. To prepare seed potatoes for planting, first slice off and discard the end of each potato on which there is a cluster of eyes—little depressions from which sprouts will rise (other eyes are scattered, not clustered, over the rest of the potato). Then cut the potatoes into blocks about the size of small eggs. Each block should weigh 1 to 2 ounces and contain one to three eyes. Spread the pieces out in a single layer in a well-ventilated and well-lighted place for four or five days before planting. To minimize loss from rot, dust the pieces with captan. Prepare the soil by digging a flat-bottomed furrow 6 to 8 inches wide and 3 to 4 inches deep; for more than one row, space the furrows 2½ to 3 feet apart. Scatter 5-10-5 fertilizer along the bottom of the furrow at the rate of 2½ pounds to 25 feet of row; mix the fertilizer with the soil. Plant the pieces of potato in the furrow, eyes up, at 12- to 15-inch intervals, and cover them with 3 inches of soil. Potatoes form close to the surface of the soil and are easily damaged by cultivation. If it becomes necessary to cultivate, penetrate the soil no deeper than 1 inch. When the plants become 8 to 10 inches tall, use a hoe to pile the soil up around the stems to a height of 3 to 4 inches; this hilling-up with soil protects the potatoes and keeps them from turning green. Do not fertilize.

Harvest young potatoes whenever the tubers are large enough to be worth digging up—usually about the time blossoms appear, seven to eight weeks after planting. Do not dig up entire plants—instead, dig carefully around the plants and remove some of the larger tubers, leaving the smaller ones to continue to grow. Potatoes for winter storage should be dug two to three weeks after the tops have died to the ground. Dig potatoes in dry weather, and take care not to bruise the skin. Dry potatoes for winter storage for two to three hours, then store them in a cool dark place, preferably at 40° to 45°, to keep them from turning green and developing an unpleasant flavor.

PEPPER
Top left and right: sweet pepper and hot cherry pepper
Bottom: hot long pepper

For climate zones and frost dates, see maps, pages 148-149.

103

POTATO
brown-skinned and red-skinned types

PUMPKIN

Before cooking, cut out all eyes and any greenish or spoiled parts. Potato tops, sprouts and green sections contain the poisonous substance solanine.

POTATO, SWEET See Sweet Potato

PUMPKIN
Cucurbita pepo

Pumpkins take a lot of space, but many gardeners plant one or two clumps, or hills, of them for fall decoration and pies. Some save space by sowing a few seeds in a row of corn, combining two plants in the same area. Others train the vines to grow on supports. Still others plant bush pumpkins, only 3 feet across. A fine bush variety is Cinderella, bearing three to six 10-inch round fruit weighing 7 pounds. Good vine varieties are Spookie, 6- to 7-inch round fruit, 4 to 5 pounds; Jack O'Lantern, oval fruit, 9 inches tall and 7 to 8 inches across, 10 to 15 pounds; and Big Max, 20 inches or more across and more than 100 pounds.

HOW TO GROW. Pumpkins are unusual among vegetables in that they grow well in light shade as well as in full sunshine. They do best in soil with a pH of 5.5 to 7.5. Sow seeds outdoors when night temperatures are expected to stay above 55° for about four months. For vine and bush varieties, prepare hills by digging holes 1 foot deep and 2 feet across; space the holes for vine types 8 to 10 feet apart, for bush types 4 feet apart. Put into each hole a bushel of compost or well-rotted cow manure, then return the soil, forming a mound about 4 inches high. Plant four to six seeds 1 inch deep in a circle on top of each hill. When seedlings appear, cut off all but one or two of the strongest plants. From late summer on, remove any blossoms or new fruit to channel the plants' energies toward the fruit that have already formed. For a huge exhibition pumpkin, allow one fruit to develop on each vine, and water heavily. To train vine varieties to grow upward, follow the directions on page 30. To grow pumpkins with corn, sow single seeds every 10 feet or so in the rows of corn. Harvest pumpkins when leaves die and the fruit become a rich orange, about four months after sowing. Cut pumpkins from the vine with pruning shears, leaving about 3 inches of stem on the fruit; pumpkins decay quickly if the stems are broken rather than cut. After harvesting, set pumpkins in the sun for a week or two to harden the outer skins, then store them in a cool dry place.

R

RADISH
Raphanus sativus

Most commercial radishes are red, but there are white, red-and-white and even black ones. There are two main types: ordinary ones, small and quick maturing, and winter ones, larger and more pungent. Both the flesh and skin of ordinary radishes are edible, but the skin of winter radishes should be removed to expose the white inner tissue. Fine ordinary varieties are Burpee White, white, round, 1 inch across; Cherry Belle, red, round, ¾ inch across; French Breakfast, red, oval, 1¾ inches long; Sparkler, red and white, round, 1¼ inches across; and White Icicle, white, 3 to 5 inches long, 1 inch across. Good winter types are Round Black Spanish, black, round, 3 to 4 inches across; and White Chinese, also called Celestial, white, 6 to 8 inches long, 2 to 3 inches across. A 12-foot row yields about 6 pounds over a week.

HOW TO GROW. Radishes grow best in soil with a pH of 6.0 to 7.0. They grow fast; fertilizer applied after planting will not reach roots, so prepare the soil before sowing: dig about 2 inches of compost or 4 inches of cow manure into a strip 12 to 18 inches wide and 8 inches deep, then add 1 pound of 10-10-10 fertilizer to each 10 feet of row. In most of the U.S. and Canada, where frost is expected in winter, sow seeds of ordinary varieties in early spring when the soil can be worked and continue to plant every 10 days until a month before maximum daytime temperatures are expected to average over 80°. In late summer, when maximum daytime temperatures average below 80°, start successive plantings of ordinary radishes again and continue until night temperatures drop to about 40°. Sow winter radishes once, about two months before minimum night temperatures average below 20°; they can be harvested in fall for storing over winter. In regions where winter temperatures rarely fall below 30°, start successive plantings of both types in fall; make the final planting of ordinary radishes a month before maximum daytime temperatures average above 80°, and make the final planting of winter radishes a month earlier.

Sow seeds ½ inch deep. Space ordinary varieties ½ inch apart in rows 4 to 6 inches apart. When seedlings are 1 to 2 inches tall, thin them to stand 1 inch apart. When sowing winter varieties, group three or four seeds in a spot; set each group 4 to 6 inches apart in rows 12 inches apart. When seedlings are 1 to 2 inches tall, cut off all but the strongest in each group. Keep the soil moist. Harvest radishes when the diameters of the roots reach the size listed for each variety, before they become tough and woody. Pull up ordinary varieties 25 to 30 days after sowing, winter varieties after 60 days.

RHUBARB
Rheum rhaponticum

Rhubarb is grown for its stout stalks, 18 to 36 inches long, which are bright red or green. The stalks are topped by dark green leaves, 12 inches or more across, that are poisonous. Rhubarb is a cold-resistant perennial that thrives anywhere in the U.S. and southern Canada where maximum daytime temperatures average no higher than 90°; it does not grow well in Florida or on the Gulf Coast. Grow crops from roots of named varieties, rather than from seeds. Most varieties send up edible stalks and 3- to 4-foot-long seedstalks, which steal strength from the roots unless cut off at ground level before the seeds mature. Fine varieties are Chipman's Canada Red and Honeyred, which do not send up seedstalks; MacDonald and Valentine; and Cherry, good for the mild winters on the West Coast. A plant yields 6 to 8 pounds over six weeks.

HOW TO GROW. Rhubarb grows best in soil with a pH of 5.5 to 7.0. Buy roots in early spring when the soil can be worked, planting them where they can grow undisturbed for many years. Prepare the soil by digging a hole about 2 feet deep and 2 feet across for each plant, spacing the plants 3 feet apart; discard the subsoil. Place a 6- to 8-inch layer of compost or well-rotted cow manure in the bottom of each hole, then fill it with a mixture of equal parts of compost or cow manure and topsoil. Set the plants into the holes so that the tops of the roots (where the buds are located) lie 3 to 4 inches below the surface of the soil. Each fall spread a 3-inch mulch of cow manure or compost over the root tops. Wait two years after planting before starting to pull stalks for eating. From then on, harvest in spring only the leafstalks that are 1 inch or more in diameter; the smaller ones will continue to grow and will die down to the ground in winter, rebuilding the plants' strength for another year. Cut off seedstalks at ground level as soon as they appear. Harvest rhubarb by grasping each stalk near its base and giving it a sideward twisting tug; the stalk will separate cleanly from the top of the roots.

RADISH
Clockwise from top left: White Icicle, Cherry Belle, French Breakfast, Burpee White, Sparkler

For climate zones and frost dates, see maps, pages 148-149.

RHUBARB

Rhubarb will produce heavy crops for many years, but when the stalks become small, dig up and divide the plants. Set the divided plants into the garden in a new place as if they were young plants bought at a nursery.

RUTABAGA See Turnip

S

SCALLION See Leek, Onion

SPINACH
Spinacia oleracea

Spinach forms a rosette of foliage 8 to 10 inches across. It comes in two types—smooth-leaved and crinkle-leaved (or savoy). The plants require cool weather and must be grown in spring or fall. In hot weather, they bolt—that is, they blossom and form seeds instead of developing into useful vegetables. America (savoy) and Viking (smooth) are good for spring planting. Hybrid No. 7 and Winter Bloomsdale (both savoy) can be planted in spring or fall. A 10-foot row yields about 3 pounds over two weeks.

HOW TO GROW. Spinach grows best in soil with a pH of 6.0 to 7.5. In most of the U.S. and southern Canada, where frost is expected in winter, sow seeds in early spring as soon as the soil can be worked and continue sowing every 10 days until six weeks before maximum daytime temperatures average over 75°. In late summer, when daytime temperatures average below 75°, start successive plantings again until six weeks before minimum night temperatures average less than 20°. In regions where winter temperatures rarely fall below 25°, start successive plantings in fall for harvesting in winter and spring; make the final planting six weeks before maximum daytime temperatures average 75°. Sow seeds ½ inch deep in rows 12 to 18 inches apart. When seedlings appear, thin to stand about 3 inches apart. When plants become large enough to touch one another, pull up every other plant (the pulled plants are edible). After this thinning, scatter 10-10-10 fertilizer around the plants at the rate of 3 ounces to 10 feet of row. Spinach is ready for harvesting when the largest leaves are 6 to 8 inches long, about six weeks after planting. To harvest, cut the entire plant off at the soil surface.

Always use fresh spinach seeds each year.

SQUASH
Summer squash *(Cucurbita pepo melopepo)*, winter squash *(C. maxima* and *C. moschata)*

Squashes, some growing on vines, others as bushes, form two main groups—summer squashes and winter squashes. All are grown in clumps on mounds, or hills, of soil.

Summer squashes have whitish or yellow flesh, and are picked in summer while immature. Excellent bush types with a spread of 3 to 4 feet are Early Golden Summer Crookneck, yellow; St. Pat Scallop Hybrid, light green; Burpee Hybrid Zucchini, pale green; Chefini Hybrid straightneck, dark green; Hybrid Cocozelle straightneck, striped green; Seneca Butterbar straightneck, yellow; and Royal Acorn, dark green (Royal Acorn is edible in summer when immature, but it is also used as a winter squash when ripe). Two hills yield about 16 pounds over eight weeks and will continue to produce until frost comes if all the fruits are harvested before they mature.

Winter squashes all have orange flesh. All the varieties listed are good for winter storage. Most are trailing vines spreading 10 to 15 feet or more, but there are also bush types spreading 3 to 4 feet as well as a semibush type spreading 5 to 6 feet. There are several good vine types. Blue Hubbard has pear-shaped blue-green fruit 15 to 20

SPINACH

inches long and weighing 25 to 30 pounds. True Hubbard has pear-shaped green fruit, 12 inches long and 10 inches across, weighing about 12 pounds. Buttercup has dark green fruit with silver stripes; each fruit is 4½ inches high and 6½ inches across and weighs 4 to 5 pounds. Waltham Butternut has creamy tan fruit, 9 inches long and 3 inches across, that weighs 3 to 4 pounds. A splendid bush type is Gold Nugget, bright orange, 3 inches high and 4 to 5 inches across, and weighing about 2 pounds. A semibush type is Bush Ebony, acorn-shaped dark green fruit 5 inches long, weighing 1 to 1½ pounds. One hill of vine types of winter squash yields about 30 pounds, of bush types about 5 pounds, of semibush types 10 pounds. Pick winter squash all at once when the leaves turn brown.

HOW TO GROW. Summer squash needs full sun; winter squash does best in full sun but will tolerate light shade. Both grow best in soil with a pH of 6.0 to 7.5. To prepare a hill for squash, dig a hole about 18 inches wide and equally deep. Put 3 to 4 inches of compost or well-rotted cow manure in the bottom of the hole, then fill the hole with a mixture of 3 parts soil (use the soil that was removed from the hole) and 1 part compost or cow manure, forming a mound about 4 inches high. Space hills for bush squash 4 to 5 feet apart, for vine squash about 10 feet apart, for semibush squash 5 to 6 feet apart.

Sow squash outdoors when night temperatures no longer fall below 55°. Set six to eight seeds 1 inch deep, evenly spaced on each hill. When plants become 3 inches tall, cut off all but the two strongest. When the plants begin to crawl along the ground, scatter ⅓ cup of 5-10-5 fertilizer around each one; mulch with 6 inches of straw or old hay.

Pick summer squashes when fruits are tender and easily punctured. The first fruits are ready about 50 days after sowing. Pick St. Pat Scallop Hybrid when the fruit are 2 to 3 inches in diameter; pick all other varieties when they are 6 to 8 inches long. All fruits, whether eaten or not, should be picked when they reach this size; if the fruits mature, the vines cease to produce.

Let winter squashes mature fully on the vines until their skins are extremely hard. True Hubbard matures about 115 days after sowing, Blue Hubbard and Buttercup about 100 days, Gold Nugget about 95 days, Waltham Butternut about 110 days. To pick winter squashes, cut them from the vine, leaving a 2- to 3-inch stem on each squash. Let the squashes cure in the sun for a week or more, then store them in a cool dry place over winter.

SWEET POTATO
Ipomoea batatas

Sweet potatoes, trailing vines grown for their nutritious roots, need night temperatures of at least 60° (they prefer 70° to 80°) for about 150 days. Therefore most are grown in southern areas, but satisfactory crops will also grow in southern New York, southern Michigan, Wisconsin and the Pacific Northwest if planting is timed carefully.

Sweet potatoes may have yellow, orange, dark red or brown roots, and white, yellow or orange flesh. There are two types—those whose flesh is dry and mealy when cooked (excellent varieties are Orange Jersey, Nemagold and Nugget) and those that are moist-fleshed (recommended varieties are Goldrush, Centennial and Porto Rico). In some parts of the South, moist-fleshed varieties are called yams, a name also used for the unrelated plant *Dioscorea*. Twenty-five feet of row yields about 25 pounds.

HOW TO GROW. Sweet potatoes grow best in light, sandy shallow soil with a pH of 5.5 to 6.5; deep rich soils produce too much foliage and stringy roots. Once the plants are established, they tolerate dry soil.

For climate zones and frost dates, see maps, pages 148-149.

SUMMER SQUASH
Clockwise from top left: crookneck, scallop, acorn, straightneck and zucchini types

WINTER SQUASH
Clockwise from top: True Hubbard, Butternut, Gold Nugget, Buttercup

Sweet potatoes are grown from sprouts, or slips, that the sweet potatoes send out. Buy slips from a nurseryman in spring, or buy some sweet potatoes and start slips yourself. To produce slips, lay the sweet potatoes on their sides in a hotbed about a month before night temperatures can be depended upon to stay above 60°. Cover the sweet potatoes with 2 inches of moist sand and keep the hotbed between 75° and 85°. When sprouts develop, give them a twisting tug (do not cut them) and they will slip right off.

To prepare the soil, dig it 4 to 6 inches deep. Apply a 2-foot band of 5-10-10 fertilizer in rows 3 to 4 feet apart; spread it at the rate of 4 pounds to 50 feet of row, and mix into the soil. Push the soil in each row into a ridge 4 to 8 inches high and about 12 inches wide. Plant the slips on the ridge, spacing them 12 to 15 inches apart; set them so that about 4 inches of each slip is buried in the soil. One or two leaves should show above the soil; bury any others.

Pull out weeds or cultivate shallowly until the vines carpet the soil. Do not add supplementary fertilizer. Dig up sweet potatoes in late fall in frost-free regions, elsewhere as soon as the tops of the plants become blackened by the first fall frost. Handle the potatoes like eggs—the skins are tender and bruise easily, and any damage may cause decay in storage. Let the sweet potatoes dry for two or three hours, then spread them out in baskets lined with newspaper and put them in a dry place where the temperature will remain about 80° for 10 days to two weeks. Gradually reduce the temperature to 50° to 55° by ventilating the curing area, but continue to keep it dry. Sweet potatoes keep about 10 weeks.

SWEDE TURNIP See Turnip
SWISS CHARD See Chard

T

TOMATO
Lycopersicum esculentum

Tomatoes, the most popular vegetable for the home garden, are divided into three main types. Small compact plants with stems only 12 to 18 inches long, called determinate, stop bearing once they reach their full size. Slightly larger plants, called semideterminate, cease production when their stems become 18 to 24 inches long. The third type, known as indeterminate, consists of wide-ranging vines that grow and bear indefinitely unless they are killed by frost or disease; indeterminate plants are the only tomato plants suitable for staking.

Modern varieties have a built-in resistance to verticillium wilt and fusarium wilt, diseases that can wipe out a crop and for which there are no effective sprays; several varieties are also resistant to nematodes, microscopic pests that eat and weaken tomato plants. The resistance of each variety listed below is indicated by the letter "V" for verticillium wilt, "F" for fusarium wilt or "N" for nematode. Several nonresistant varieties are recommended because many gardeners find them worth growing for their flavor, color, size or vigor, even at risk of loss.

Except as noted, all the varieties recommended are large-fruited red tomatoes. Each variety is listed with its type, the average time required for the first fruit to ripen after 6- to 8-inch seedlings are planted outdoors, and the letter indicating its pest and disease resistance. In southern Canada, where the frost-free period is short and maximum summer daytime temperatures average less than 90°, recommended varieties are Burpee's VF Hybrid, indeterminate, 72 days, VF; Spring Giant, semideterminate, 65 days, VF; and Springset, determinate, 67 days, VF. In the East and

SWEET POTATO

Northeast, where the frost-free period is longer than it is north of the Canadian border but maximum summer daytime temperatures still average less than 90°, recommended varieties are Beefeater, indeterminate, 75 days, VFN; Better Boy, indeterminate, 70 days, VFN; and Spring Giant, semideterminate, 65 days, VF. In the Midwest, where winters are extremely cold but maximum summer daytime temperatures average over 90°, recommended varieties are Better Boy, indeterminate, 70 days, VFN; Bonus, semideterminate, 75 days, VFN; and Campbell 1327, semideterminate, 69 days, VF. In the Northwest, where maximum summer daytime temperatures average less than 90° and cloudiness and mist are common, recommended varieties are Beefeater, indeterminate, 75 days, VFN; Terrific, indeterminate, 70 days, VFN; and Vineripe, indeterminate, 80 days, VFN. In the Southwest, where the climate is arid and maximum summer daytime temperatures average over 90°, recommended varieties are Beefeater, indeterminate, 75 days, VFN; Better Boy, indeterminate, 70 days, VFN; and Spring Giant, semideterminate, 65 days, VF. In the South, where summers are long and hot, and maximum summer daytime temperatures average over 90°, recommended varieties are Better Boy, indeterminate, 70 days, VFN; Manalucie, indeterminate, 82 days, F; and Tropic, indeterminate, 82 days, VF.

Several varieties are excellent for special purposes and can be grown in all regions. Roma (determinate, 75 days, VF), an oval tomato 2½ to 3 inches long and 1½ inches in diameter, is extensively used to make tomato paste. Small Fry (determinate, 52 days, VFN) bears great quantities of red fruit about 1 inch in diameter and can easily be grown as a pot plant on a sunny patio. Sunray (indeterminate, 72 days, F) is grown for its 2½- to 3-inch golden-orange fruit. Three nonresistant varieties are Ponderosa (indeterminate, 83 days), whose pink fruit, up to 6 inches across and weighing as much as 1½ pounds each, is excellent for slicing; and Yellow Pear and Yellow Plum (both indeterminate, 70 days), two vigorous plants with pear- and plum-shaped yellow fruit about 2 inches long.

Determinate plants yield about 8 pounds each, semideterminate 10 pounds, and indeterminate 8 pounds if staked, 15 pounds if unstaked. A cold, rainy growing season will cause delay in ripening and reduce the yield.

HOW TO GROW. Tomatoes grow best in soil with a pH of 5.5 to 7.5. For early crops in regions with winter frost, start plants indoors or in a hotbed five to seven weeks before night temperatures are expected to stay above 60°; set the seeds ⅛ inch deep. When seedlings become about 1 inch tall, transplant them to individual 3- to 4-inch pots. Move the plants to the garden when night temperatures are expected to remain above 60°. To prepare the ground, scatter a 1-foot-wide band of 5-10-5 fertilizer at a rate of 1½ pounds to every 25 feet of row; dig the fertilizer in thoroughly, and set the plants 2 to 3 feet apart in rows 4 feet apart. For later crops, sow seeds directly outdoors when night temperatures can be depended upon to stay above 60°. Group three or four seeds in a spot, setting each group ½ inch deep and 2 to 3 feet apart in rows 4 feet apart; when the seedlings become 1 inch tall, cut off all but the strongest one. In frost-free regions, sow seeds outdoors in early fall for harvesting in winter and spring. Tomatoes may also be started from small purchased plants; set them in the garden when seeds or seedlings should be planted outdoors. Indeterminate tomatoes may be allowed to ramble on the ground, or they may be staked (drawings, page 29); other types are not supported.

Because tomato plants are susceptible to a disease called blossom-end rot, a mulch of grass clippings or old hay is es-

TOMATO
Left: large-fruited red type   *Right:* small-fruited red type

TOMATO
Top: large-fruited orange type
Bottom: yellow and red pear-shaped types

For climate zones and frost dates, see maps, pages 148-149.

TURNIP

RUTABAGA

pecially advisable. To prevent cutworm injury, cover each plant with a paper cup that has had its bottom removed. Once a month scatter a handful (about ⅓ cup) of 5-10-5 fertilizer around each plant in a circle extending at least 2 feet from the stem in all directions. A staked plant must be trained to one or two stems; to remove side stems that grow between each leaf and the main stem, wait until they are 1 or 2 inches long, then pull them down sharply—they will snap right out. To assure a crop early in the season, when temperatures are cool, spray the blossoms with a hormone compound designed to stimulate the formation of fruit. Tomatoes are ready to be harvested when they have developed their full color; to pick them, gently lift each tomato until the stem snaps.

TREE CABBAGE See Collard

TURNIP and RUTABAGA
Turnip *(Brassica rapa);* rutabaga, also called Swede turnip *(B. napobrassica)*

Turnips and rutabagas are cool-weather plants grown for their tender, crisp roots. The roots of the two plants look alike except that those of turnips are about 2 inches across and have white flesh, while those of rutabagas become 4 to 5 inches across and have white or yellow flesh. Both are excellent cooked. The leaves, or greens, of turnips are also good cooked, but the tops of rutabagas should be discarded. Choice varieties of turnips are Just Right, Purple-Top White Globe and Tokyo Cross; of rutabagas, American Purple-Top (yellow flesh), Improved Long Island (yellow flesh) and Macomber (white flesh). A 15-foot row of turnips or rutabagas yields about 15 pounds over a period of six weeks.

HOW TO GROW. Turnips and rutabagas grow best in soil with a pH of 5.5 to 7.0. In most of the U.S. and southern Canada, where frost is expected in winter, sow turnip seeds in early spring as soon as the soil can be worked and continue to sow every two weeks until five weeks before maximum daytime temperatures are expected to average above 80°. In late summer, when maximum daytime temperatures average below 80°, start successive plantings again until three months before minimum night temperatures average less than 20°. In regions where winter temperatures rarely fall below 25°, start successive plantings in late summer and continue until five weeks before maximum daytime temperatures average 80°.

Plant rutabagas annually. In areas with winter frost, sow seeds in early summer. In frost-free areas, sow in fall when the maximum daytime temperatures no longer are expected to average 80° or more.

To sow seeds, group three or four in a spot, setting each group ½ inch deep. Space groups of turnip seeds 2 to 4 inches apart in rows 12 to 15 inches apart. Space groups of rutabaga seeds 6 to 8 inches apart in rows 15 to 24 inches apart. When seedlings become 1 inch tall, cut off all but the strongest plant in each group. When the plants are 4 inches tall, scatter a 6- to 8-inch band of 5-10-5 fertilizer along each side of the row at the rate of 3 ounces to 10 feet of row. Turnips are ready to be pulled up when the roots are 2 inches in diameter—in the case of Just Right, 35 to 40 days after seeds are sown; Purple-Top White Globe, 45 to 55 days; and Tokyo Cross, 35 days. All rutabaga varieties can be pulled about three months after seeds are sown, or they can be mulched and kept in the ground until needed. To harvest turnips and rutabagas, grasp the tops and pull them up.

TURNIP-ROOTED CELERY See Celery

Fruits

A

APPLE
Malus pumila

Apples, the most familiar of all temperate-climate fruits, are grown throughout the world in innumerable varieties. Most are 2 to 4 inches in diameter, but vary in color from bright red to yellow and green, and many are blended and striped rather than a solid color. The fragrant blossoms are about 1 to 1½ inches in diameter and change from pink to white as they open in spring.

Until recently nearly all apple trees have been standard-sized trees that may reach a height and spread of 25 feet, require 5 to 10 years to produce the first fruit and yield too much fruit for the average family to consume. Fortunately for home gardeners, apple trees are now available in more practical dwarf and semidwarf forms that bear fruit as large as, if not larger than, that of standard trees and produce it more quickly. Dwarf trees, only 6 to 8 feet tall, will bear fruit two or three years after planting, and semidwarfs, 12 to 15 feet tall, may start to bear three or four years after planting. The small size of these new types permits several varieties to be grown in an area that a single full-sized tree would occupy; an even greater diversity of types can be achieved by grafting several varieties onto each small tree. Where space is very limited, dwarf trees can be espaliered against a wall or fence *(drawings, page 55)* or even trained to grow as a hedge.

Some varieties of apples bear fruit in early summer, others in late summer and still others in fall. Among these are types suited for areas as cold as Zone 4 or as warm as Zone 8, but most types do best in Zones 5-7. Some, however, do better in one region within these climate zones than in another. Dependable varieties for the eastern part of the country in Zones 5-7 are Early McIntosh, Gravenstein, Lodi, Starkspur Earliblaze and Stark SummerGlo, all early summer; Cortland, Macoun, McIntosh, Red Delicious and Wealthy, all late summer; and Jonathan, Golden Delicious, Grimes Golden, Northern Spy and Stayman Winesap, all late fall. In Zones 5-7 in the Northwest recommended varieties are Gravenstein and Lodi, both early summer; Starkrimson, Starkspur and strains of red-fruited Delicious, late summer; and Golden Delicious, Jonathan and Rome Beauty, all late fall. In Zones 5-7 of northern California the varieties most frequently grown are Red June, early summer; Gravenstein, midsummer; Jonathan, early fall; Golden Delicious, fall; Rome Beauty and Yellow Newtown, late fall. The varieties that do best in the southern edge of the apple-growing area (Zone 8) are Summer Champion, late summer; Winter Banana, fall; and Tropical Beauty, late fall. Varieties especially adapted to cold regions (Zone 4) are Anoka and Wealthy, early fall; Harlson and Secor, late fall. In addition to these varieties many excellent old-fashioned apples *(pages 46-48)* are available from specialists as young trees or as young branches for grafting. Check with your local nurseryman for the varieties best suited to your area. Trees of at least two varieties should be planted within 50 feet of one another, because pollination of one variety by the pollen from another is usually required for the trees to bear. Apple trees may bear crops for 30 to 50 years. A standard tree may yield 15 to 20 bushels of fruit annually, while a dwarf tree bears 1 to 2 bushels and a semidwarf 5 to 10 bushels.

HOW TO GROW. Apple trees grow best in soil with a pH of 5.5 to 6.5. Buy vigorous one- or two-year-old trees about 3 to 5 feet tall and plant them in late fall in Zones 6-8; plant in early spring in Zones 4 and 5. To reduce frost dam-

APPLE
Left: Golden Delicious   *Center:* Rome Beauty
Right: Red Delicious

For climate zones and frost dates, see maps, pages 148-149.

APPLE
Top: Stayman Winesap *Center:* Jonathan *Bottom:* Lodi

APPLE
Top left and right: McIntosh and Starkspur Earliblaze
Bottom: Northern Spy

age, it is best to plant apple trees on a slope so that cool air can flow away to a lower elevation. Plant standard apple trees 40 feet apart; semidwarf trees should be spaced 20 feet apart and dwarf varieties of apples 10 to 12 feet apart. Cut off all but three or four of the tree's strongest, best-placed main branches; these should be well distributed, about 6 to 12 inches apart, and form angles greater than 45 degrees with the trunk. Prune these remaining branches to 6- to 8-inch stubs.

To judge whether a standard tree is receiving sufficient fertilizer, check the leaf color and texture: thick, broad, dark green leaves indicate the presence of sufficient fertilizer. If leaves are pale or yellowish green, scatter 10-10-10 fertilizer beneath the branches in early spring at the rate of 1 pound per 3½ inches of trunk diameter for a standard tree; use a total of ¼ pound for a dwarf or semidwarf tree 3 to 5 feet tall, and use a total of ½ pound for a dwarf or semidwarf tree 6 to 8 feet tall. It is better to err by giving too little than too much fertilizer; overfeeding promotes many leaves but few apples, and it also reduces a tree's resistance to winter cold.

When apple trees begin to bear fruit, it is important that they carry only as much fruit as they can support. Part of the problem of overproduction is taken care of naturally by what is known as June drop, the falling of some fruit in early summer. Thin out the remaining fruit when they are about one third grown so that the apples are about 6 to 7 inches apart; there should be about 30 to 40 healthy leaves for each apple. If apple trees bear little or no fruit, the reason may be fruit overproduction during the previous year, a late frost that killed the flower buds, overfertilization or lack of pollination because only one variety has been planted.

One of the major problems in apple growing has always been the control of insects and diseases: about two dozen types of insects attack apple trees in all parts of the country, and the apple tree is vulnerable to several fungus and bacterial diseases. To be assured of worm-free fruit, spray trees with an all-purpose spray as described on page 151.

Apples should be picked by hand when ripe but still firm. Raise the fruit to one side with your fingers and twist your wrist; if ripe, the fruit will release easily. Do not shake or yank it off; if you do, the fruit spur may come off along with the fruit, preventing further fruit production on the spur, or the stem may pull from the fruit, breaking the skin and opening a site for rot.

APRICOT
Prunus armeniaca

The familiar apricot, believed to have originated in China, has been cultivated for many thousands of years. A standard tree grows about 20 feet tall, spreading in a vase shape to a diameter of 25 to 30 feet; dwarf varieties grow about 8 feet tall, spreading to 10 feet. Apricots have attractive pink flowers that open so early in the spring that they are sometimes nipped by frost. They are followed in late summer by 1½- to 2-inch fruit with orange, very sweet flesh. The fruit are most tasty when allowed to ripen on the tree; when ripe they are plump, fairly firm and a uniform golden-yellow color.

Apricots grow in Zones 4-8. Two of the best varieties adaptable to many zones are Moorpark and Early Golden. In California favorite varieties are Royal, Blenheim and Tilton, and in Zone 4, Moongold and Sungold are recommended. All apricot trees but Moongold and Sungold will bear fruit if planted alone. Moongold and Sungold, however, should be planted together, for pollination of one variety by pollen from the other is required if the trees are to bear fruit. Apricot trees may survive and bear for 35 years or more. A full-grown standard tree yields about 3 to 4 bushels of fruit yearly; a dwarf yields about 3 bushels.

HOW TO GROW. Apricots grow best in soil with a pH of 6.0 to 7.0. For fruit within three or four years, buy one-year-old trees 3 to 5 feet tall. Plant them, if possible, on a north-facing slope; there they will not blossom so early that they will be caught by frost, and the slope allows cold air to flow away. Plant standard-sized trees 30 feet apart, dwarf trees 10 to 15 feet apart, as early in the spring as the ground can be worked. Cut off all but the three best-placed branches. They should be 6 to 12 inches apart, face in different directions and form angles greater than 45 degrees with the trunk. Cut them back to 2 to 4 inches. In the future, pruning should be restricted to removing deadwood and overcrowded branches.

To judge whether an apricot tree is receiving sufficient fertilizer, check the appearance of the foliage. If the leaf color in summer is pale or yellow-green, feed each tree with 10-10-10 fertilizer early the second spring after planting at the rate of ¼ pound for each year of the tree's age. Scatter it on the ground beneath the branches.

When apricot trees begin to bear fruit, it is important that they carry only as much fruit as they can support: if they are overproductive, all of their energy will go into fruit production instead of flower-bud formation and food storage for the following year's fruit. When the apricots are about ¾ inch in diameter, thin them by pulling off excess fruit; those left should be about 3 inches apart.

Apricots should be hand-picked when they reach the golden-yellow color and firm softness of maturity.

B

BLACKBERRY
Blackberry, including boysenberry, loganberry and youngberry *(Rubus)*

Freshly picked fully ripe blackberries, boysenberries, loganberries and youngberries have delectable flavors seldom experienced except by those who grow their own plants —the fresh berries are so soft that they cannot be shipped, and nearly all the berries grown commercially are frozen or canned, or made into jam or wine.

Despite their many names, botanists consider all of these berries to be varieties of blackberries. The additional names were coined by an enterprising nurseryman years ago and, in the process, confused generations of gardeners who sought out these "wonder" berries, thinking them to be of some exotic species. They are all hybrids of blackberries and red raspberries and have reddish black fruit, instead of the black fruit of the ordinary blackberry.

Individual blackberries are 1 to 2 inches long and are composed of many united segments, each of which carries a small seed. Some plants are stiffly erect bushes growing 4 to 5 feet tall, while others are trailing vines that must be tied to a support to keep the fruit off the ground. Both types are alike, however, in that their canes grow one year, then bear fruit and die the second year. Each summer new canes are produced; these will bear fruit the following year. Nearly all blackberries are notorious for their thorny canes, although a few thornless varieties exist; they are much easier to work among, but generally bear small crops.

Bush types of one variety or another grow in many climates. For black-fruited bush-type berries, Darrow and Eldorado are excellent varieties in Zones 5-8; Lawton is best suited to Zones 8-10 and Snyder to Zones 3-5.

Most vine types are susceptible to frost damage and can be grown only in Zones 7-10. Some good varieties of black-fruited vine-type berries are Brainerd and Lucretia, suited

APRICOT
Sungold

For climate zones and frost dates, see maps, pages 148-149.

BLACKBERRY
Darrow

to eastern parts of Zones 7-10; Marion, Zones 7-9 in Washington and Oregon; and Olallie, Zones 7-9 in California. For red-fruited vine-type blackberries, recommended varieties are Boysen, which can be grown throughout Zones 7-9; Cascade, Zones 7-9 in Washington and Oregon; Logan, Zones 7-9 on the West Coast; and Young, Zones 7-9 on the West and Gulf Coasts.

Blackberry plants are long-lived, usually bearing fruit for as long as 20 years, and are extremely prolific: a single plant may produce several quarts of berries during the midsummer fruiting period.

HOW TO GROW. Blackberries grow best in a moist but well-drained soil of pH 5.5 to 7.5 that has been supplemented with compost or manure. From Zone 5 north, plant blackberries in the early spring as soon as the ground can be worked; from Zone 6 south, set the plants out in the fall, winter or spring.

The culture of bush and vine types differs considerably, primarily because of their habits of growth. Set bush-type plants 4 to 5 feet apart in the row and space rows 6 feet apart. Cut back the canes of newly planted bushes to 6 inches from the ground. During the summer, mulch with 4 to 6 inches of old hay or 2 to 3 inches of wood chips, sawdust or ground bark. When the new shoots that grow during the summer reach a height of about 3 feet, cut off the tips of the stems to force the development of side branches. Late in the following spring after flower buds become visible, remove weak canes and thin others to stand about 10 inches apart. Then reduce the length of the side branches to about 18 inches; the buds on the remainder of the side branches send out stems that produce white flowers that are followed by fruit in midsummer. While the previous season's growth is producing fruit, new stems are growing for another season's crop; cut off the tips of these canes. Late in the summer after the fruiting season has passed, cut out and destroy all canes that have borne fruit.

Vine-type blackberries should be planted 6 feet apart in rows 8 feet apart; cut back newly set plants to 6 inches from the ground and mulch as described above. During the first season allow the vines to creep along the ground until they are 8 feet long; then cut off the tips to force the development of side branches (drawings, page 51). During the summer or early in the following spring, install posts at 15-foot intervals along the rows and nail two strands of galvanized-iron wire from post to post at different heights. The vines can be tied to the wires during the summer where winters are mild. However, frost damage is less likely if the canes are allowed to lie on the ground through the winter in Zone 7; it is also a good idea to cover them with mulch. In spring cut off all but 16 canes from each plant, and tie them in bunches of four to the wires. Cut back side branches to 12 inches, and as with bush-type plants, remove fruit-bearing canes after the season has passed.

Feed both bush- and vine-type plants each spring by scattering 5-10-5 fertilizer around them at the rate of 1 cup per 16 feet of row.

Blackberries should be harvested in midsummer when the berries are so ripe that they drop off at the slightest touch. Unripe berries or those allowed to sit in the sun after picking have a bitter taste. To avoid loss of fruit to birds, cover the plants as they ripen with plastic netting.

Bush-type blackberries are propagated by digging up and replanting suckers, fast-growing stems that arise from the roots. Vine-type plants are propagated by tip layering: cover the tips of the canes with 2 inches of soil in midsummer. Roots and latent stem-buds will develop under the covered area during the fall and winter; cut off the tips of the canes and transplant the new plants in early spring.

BLUEBERRY

High-bush blueberry *(Vaccinium corymbosum)*, rabbiteye blueberry *(V. ashei)*

Blueberries are ornamental as well as useful: the rounded bushes may grow 5 to 6 feet tall and have clusters of white ¼-inch flowers in the spring, rich green foliage that turns deep red in the fall and abundant crops of sweet blue berries about ½ inch in diameter in midsummer.

The best blueberry-growing areas of the country are those where the soil is moderately acid. In the eastern part of the country, the area stretches from Maine to Michigan, Wisconsin and Minnesota south to Florida and Louisiana. In the West the best area is the coastal strip west of the mountains from Washington to northern California. In the rest of the country blueberry culture may be difficult, but isolated plants can be grown in many areas by following the special culture method noted below. The most important blueberries in most of the country are the high-bush type. Excellent varieties, in their order of ripening, are Earliblue, Collins, Blueray, Bluecrop, Berkeley, Jersey, Dixi, Herbert and Coville. Of these named varieties, the ones most suited for the Northwest are Earliblue, Bluecrop, Berkeley, Jersey and Dixi. In the southern part of the country from South Carolina to Florida west to Mississippi and Louisiana, rabbiteye blueberries do better than high-bush blueberries. Good varieties, in their order of ripening, are Tifblue, Callaway, Walker and Homebell. In North Carolina varieties resistant to canker are useful; the best ones, in their order of ripening, are Angola, Wolcott, Croatan, Murphy and Scammell.

At least two varieties of blueberries should be planted, for pollination of one variety by pollen from another is usually necessary to ensure that the bushes will bear fruit. Blueberry bushes will live and bear for 35 to 40 years; an average plant bears 6 to 8 pints of berries annually, but it may bear at least twice that amount if given good care.

HOW TO GROW. Blueberries do best in acid soil with a pH of around 5.0. The soil should be supplemented with peat moss, which should make up 50 per cent of the soil mixture around the plants. From Zone 5 north, set out the blueberries in the early spring as soon as the ground can be worked; from Zone 6 south, set them out in fall, winter or spring. For berries within three years, buy plants that are 1 to 2 feet tall. Smaller plants, usually sold as one- or two-year-old plants, should be set in a nursery bed for a few years before being set into the garden. If possible, buy blueberry plants with a ball of soil around their roots, for the roots are very fine and hairlike and must be kept moist at all times. Set high-bush plants 5 to 6 feet apart in rows 8 feet apart; rabbiteye varieties should be planted 8 feet apart in rows 8 to 10 feet apart. Cut back both types about halfway at the time of planting.

Immediately after planting, mulch the soil with about 4 inches of old sawdust, wood chips or ground bark. Retain at least this much mulch at all times, increasing the depth on mature plants to as much as 8 inches.

Blueberry bushes should be pruned during their dormant season in winter or early spring. Remove thin, weak growth and old wood—the largest fruit are borne on the fresh new canes. Some varieties bear so many berries that nourishment is exhausted and the fruit are undersized. To avoid this problem, cut back the tips of canes so that only four or five fat flower buds are left on each twig. Pick off all blossoms the first two years and allow only a small crop to mature during the third season. Thereafter the crop will increase gradually each season until the plants reach maturity in 8 to 10 years.

In areas where the soil is alkaline, having a pH of 7.0

BLUEBERRY
Earliblue

For climate zones and frost dates, see maps, pages 148-149.

or above, plant blueberry bushes in containers sunk into the ground. Two very satisfactory containers can be made by cutting one 50-gallon metal drum in half, then punching three or four 2-inch holes in the bottom of each section for drainage. Sink the drums into the soil to within 1 inch of their rims and fill them with an acid type of peat moss, such as sphagnum moss, or a mixture of equal parts of sphagnum moss and sharp sand. Plant and mulch the bushes as recommended above. Despite this special planting, blueberries in alkaline areas will have pale yellowish green leaves with dark veins, a condition known as iron chlorosis, unless extra precautions are taken—ground water and even water from the garden hose may be alkaline enough to cause the chlorotic reaction. To counteract chlorosis, apply a solution of a chemical called iron chelate —1 tablespoon to a gallon of water for each plant— sprinkling it on the leaves and soil whenever the leaves lose their dark green color.

Fertilizer should be applied only if the leaves are uniformly pale, indicating a need for nourishment. Blueberries' roots are so fine they cannot assimilate strong fertilizers. The safest source of nutrients is cottonseed meal, which is nonburning and decays slowly. Use ¼ pound around young plants and ½ pound around old ones; apply it very early in the spring, scattering it beneath the branches. However, ammonium sulfate is an excellent chemical fertilizer for old plants if cautiously applied: use a maximum of ¼ pound for a plant 5 to 6 feet tall. Ordinary 10-10-10 fertilizer can also be used at the rate of 1½ ounces for each year the plant has been in the garden, up to 8 ounces.

Blueberry varieties ripen over a two-month period in midsummer in most of the country, but the season's first berries ripen in May in Florida. To get sweet blueberries, allow the fruit to remain on the plants at least one week after they turn blue; unripe berries are sour. To pick only the ripe ones, cup the berry cluster in your hand and gently roll the darkest blue berries with your thumb. If they are ripe, they will drop easily into your hand; if any pressure is needed to loosen them, they are unripe and should be left on the bush. Pick at weekly or 10-day intervals.

Birds are the greatest threat to blueberry culture, and the only defense against them is to cover each plant with black plastic netting; clip the ends together beneath the plant with clothespins. Put the net on as the berries begin to turn pink, and remove it after the fruit have been picked.

BOYSENBERRY See Blackberry

C

CHERRY

Sweet cherry *(Prunus avium);* sour cherry, also called tart cherry *(P. cerasus);* western sand cherry *(P. besseyi)*

Fully ripe sweet cherries are commonly eaten either raw or cooked, but sour and bush cherries are almost always used for cooking; however, the ripe fruit have a fine tangy taste when eaten fresh.

Sweet cherries grow in Zones 5-7 on upright trees about 25 feet tall with an equal spread. They have 1-inch white blossoms and 1-inch fruit. Popular varieties for the western part of the country are Bing, red; Black Tartarian, black; Lambert, black; Napoleon, also called Royal Ann, yellow; Van, black; and Vista, red. Lambert, Napoleon and Bing will not bear any fruit if grown alone or together; Black Tartarian, Van or Vista is needed to pollinate them so that they bear fruit. Plant at least one of the latter group with any of the others, setting trees within 50 feet of each other. In the eastern part of the country the recommended varieties, any of which will pollinate another variety on the list, are Black Tartarian, black; Schmidt's Bigarreau, black; Napoleon, yellow; Windsor, red; and Emperor Francis, red. Two of the best varieties for cold areas are Lambert and Windsor, which will pollinate one another. A sweet cherry tree usually lives 30 to 40 years and yields 2 to 3 bushels of fruit annually.

Sour cherries grow throughout Zones 4-7. A particularly good tree is Montmorency, 12 to 15 feet tall; Meteor, 10 to 12 feet tall, and North Star, 6 to 8 feet tall, also may be grown. All have 1-inch white blossoms and ¾-inch red fruit. Trees of at least two varieties should be planted, for pollination of one variety by the pollen from another will usually increase the crop. Sour cherries live 30 to 40 years and each tree yields 1 to 2 bushels of fruit annually.

Western sand cherries grow 4 to 7 feet tall and can be grown throughout Zones 3-7; they are particularly appreciated by gardeners in Zones 3 and 4, beyond the northern range of tree cherries. Their ½-inch white blossoms are followed by ½-inch tart fruit that are used in pies, jams and jellies. Excellent varieties are Hansen's Improved, purple; Black Beauty, black; Brooks, red; and Sioux, black. Bushes of at least two different varieties should be planted, for pollination of one by the pollen from another is necessary for the bushes to bear fruit. Each bush yields several quarts of fruit annually.

HOW TO GROW. Cherries grow best in soil with a pH of 6.0 to 8.0; they are not tolerant of wet soil. Buy 3- to 5-foot-tall sweet or sour trees for fruit within three to five years; buy 1½- to 3-foot bush cherries for fruit the following year. Because cherry trees and bushes start growing so early in spring, it is best to plant them in the fall or early spring as soon as the ground can be worked. Cut off all but two to four of the best-placed branches; these should be 6 to 12 inches apart, face in different directions and form angles greater than 45 degrees with the trunk. Cut these branches back to 8 to 10 inches. Further pruning should be light and limited to removing dead and weak branches or those that grow higher than desired. In the case of bush cherries, cut plants back halfway when planting. Further pruning of them is rarely necessary.

To judge whether a cherry tree is receiving sufficient fertilizer, check the color of the foliage. If leaves are pale or yellowish green, scatter a 10-10-10 fertilizer beneath the tree in the early spring at the rate of ½ pound for each year of the tree's age until it reaches mature size. Western sand cherries should be given ¼ to ½ pound of fertilizer each spring, depending upon the size of the bushes. Sweet and bush cherries are ready for harvesting in early summer, sour cherries soon afterward. Cherries keep best if picked with the stems attached. Grasp the stems and give them a twist; try not to break off the tiny spurs because fruit will be borne on them in future years.

The greatest threat to cherries is from birds, which enjoy the cherries so much that they begin to eat them as soon as they show the slightest tinge of the ripe color. One way to reduce bird damage is to spread plastic netting over the trees. This technique is very practical for bush cherries, and it can also be used on sour cherries if the trees are kept under 10 feet in height by pruning. Sweet cherries, however, are usually too large for covering, and some gardeners use a diversionary tactic: they plant a mulberry tree close by. As much as birds like sweet cherries, they like mulberries even better.

CITRUS

Orange *(Citrus sinensis)*; grapefruit *(C. paradisii)*; lemon *(C. limonia)*; lime *(C. aurantifolia)*; mandarin orange, Satsuma orange and tangerine *(C. reticulata)*; tangelo *(C. hy-*

CHERRY
Left: Montmorency   *Right:* Windsor

For climate zones and frost dates, see maps, pages 148-149.

brid: tangerine and grapefruit); tangor (*C*. hybrid: tangerine and orange); kumquat (*Fortunella* species); limequat (*C*. hybrid: lime and kumquat)

Citrus-fruit trees are available as standard trees and, in some parts of the country, in dwarf sizes as well. They have large dark leathery evergreen leaves, and on some the branches are thorny. The clusters of 1-inch flowers, white or creamy white, are exceedingly fragrant. They usually appear most abundantly in the spring, but they may open at other seasons. Only about 2 per cent of the flowers actually produce fruits, but even this small proportion is enough to ensure a large crop. The fruits appear even if the flowers are not pollinated—thus, citrus trees can be grown singly. Unpollinated flowers produce seedless fruits, pollinated ones seeded fruits. Although there are many quite different types of citrus fruits, a number of them can be grown in a small area by grafting branches of desired types onto one tree.

Most citrus trees grow only in the frost-free regions, Zones 9 and 10, but a strip about 100 miles wide along the Gulf of Mexico is suitable for growing some of the hardier citruses, such as kumquats, mandarin and Satsuma oranges, and tangerines. Citrus trees may live and bear fruits for 100 years if given proper care. Their yields vary somewhat according to variety and region.

There are three basic kinds of oranges. The navel type is 3 to 4 inches in diameter, sweet-tasting, thick-skinned and often seedless; the common, or sweet, orange is about 2¾ inches in diameter with relatively thin, tight skin, a variable number of seeds and very juicy, sweet flesh; the blood orange is about 3 inches in diameter and has reddish skin, flesh and juice. Navel oranges usually ripen in early winter; common oranges may ripen at any season, but Valencia, the most widely grown common orange, matures in the summer; blood oranges usually ripen in the spring. Standard trees grow 20 to 25 feet tall, dwarfs 8 to 10 feet. A navel orange standard tree yields about 500 pounds of fruit annually, a dwarf about 150. A common orange standard tree yields nearly 1,000 pounds annually, a dwarf about 250 pounds. A blood orange standard tree yields 300 to 400 pounds annually, a dwarf about 150 pounds.

The recommended varieties for Arizona are Dillar, Hamlin and Marrs, all common oranges; Ruby, a blood orange; and Washington Navel. In California recommended varieties are Trovita and Valencia, common oranges; Torocco and Ruby, blood oranges; and Washington Navel. In Florida recommended varieties are Hamlin, Parson Brown, Pineapple and Valencia, common oranges; and Ruby. In Louisiana recommended varieties are Hamlin, Louisiana Sweet and Valencia, common oranges; and Ruby. And in Texas recommended varieties are Hamlin, Pineapple, Valencia and Ruby.

The lemon-yellow grapefruit, which averages 3½ to 6 inches in diameter, grows on 30- to 35-foot standard trees and 10- to 15-foot dwarfs. In Arizona recommended varieties are Marsh and Redblush; in California, Marsh and Ruby; in Florida, Duncan, Marsh and Ruby; and in Louisiana and Texas, Duncan and Ruby. The varieties Marsh and Duncan have white flesh; Ruby and Redblush have pink flesh. Marsh is unique in that even when pollinated its fruit are nearly seedless. Most varieties ripen between late fall and early spring. One standard tree yields more than 400 pounds annually, a dwarf around 150 pounds.

Standard lemon trees grow 10 to 25 feet tall, dwarfs 5 to 10 feet tall. The bright yellow fruit vary in degree of acidity and in size; most are 1½ to 2 inches in diameter and 2½ to 3½ inches long. They ripen through the year, but the greatest part of the crop matures in spring

ORANGE
Valencia

GRAPEFRUIT
Marsh

and fall. Recommended varieties for Arizona are Eureka, Lisbon and Villafranca; for California, Eureka, Lisbon, Meyer and the mild-flavored Ponderosa; for Florida, Eureka, Lisbon, Meyer and Villafranca; and for Louisiana and Texas, Meyer. A standard tree yields around 1,000 pounds of fruit annually, a dwarf 200 to 250 pounds.

Limes bear throughout the year, but more fruit ripen in early summer than in other seasons. Standard trees are 15 to 20 feet tall, dwarfs 7 to 10 feet tall. Ripe limes are yellowish green or green in color and about 1½ to 2 inches in diameter. Recommended varieties for Arizona and California are Bearss Seedless and Mexican; for Florida, Persian, also called Tahiti; and for Texas, Mexican. A standard tree yields around 1,000 pounds annually, a dwarf 200 to 250 pounds.

Mandarin and Satsuma oranges and tangerines have a delicate sweet flavor and are easy to peel because the skin is often puffy and loosely attached to the flesh. The orange fruit, which mature from fall to spring depending on the variety, are shaped like flattened globes and are 1½ to 2 inches in diameter. The standard trees grow 10 to 15 feet tall, dwarfs 5 to 8 feet tall. In Arizona and California good varieties are Clementine and Dancy tangerines, Kinnow mandarin and Owari Satsuma; in Florida and Texas, Dancy and Robinson tangerines and Owari Satsuma; and in Louisiana, Dancy and Robinson tangerines and Owari and Armstrong Early Satsuma. A standard Satsuma tree yields up to 500 pounds annually, a dwarf around 100 pounds. A standard mandarin orange or a tangerine produces about 1,000 pounds, a dwarf about 400 pounds.

Tangelos are orange-colored hybrids between the mandarin orange and grapefruit, and are between them in size. Standard trees grow 25 to 30 feet tall, dwarfs 10 to 15 feet tall. Most ripen in early spring. Recommended varieties for Arizona, Louisiana and Texas are Minneola and Orlando; for California, Minneola and Sampson; for Florida, Minneola, Orlando and Seminole. A standard tree yields about 1,000 pounds annually, a dwarf 400 to 500 pounds.

Tangors, crosses between common oranges and tangerines, are typified by the 2½- to 3-inch Temple, one of the most delicious and easy to peel of citrus fruits. Standard trees grow 10 to 12 feet tall, dwarfs 5 to 6 feet tall; fruit ripen in spring. A standard tree yields about 1,000 pounds annually, a dwarf about 400 to 500 pounds.

Kumquats bear 1- to 1½-inch orange-colored fruit that ripen in the fall; they have a mild flavor and a sweet, spicy rind that is eaten along with the flesh. Most are made into preserves. Kumquat trees are bushlike in appearance, growing 4 to 5 feet tall. Recommended varieties are Meiwa and Nagami. One tree will yield about 40 pounds of fruit a year.

Limequats, hybrids of limes and kumquats, are similar in appearance to limes and serve the same purposes, although they are not as bitter. Most fruit are about 1 inch in diameter and pale yellow when ripe. The main crop ripens during the winter, but some fruit mature throughout the year. Standard trees are only 6 to 10 feet tall, dwarfs 3 to 5 feet. In Arizona, California and Florida the recommended variety is Eustis; in Louisiana and Texas it is Lakeland. A standard tree yields about 50 pounds of fruit annually, a dwarf about 30 pounds.

HOW TO GROW. Citrus trees grow best in soil with a pH of 6.0 to 6.5. For fruit within two years, buy two-year-old trees. If bought in containers, remove them at the time of planting. Plant trees in very early spring as soon as the ground can be worked. Cut away all but three or four of the strongest branches; they should be 6 to 12 inches apart, face in different directions and form angles greater than

LEMON
Eureka

For climate zones and frost dates, see maps, pages 148-149.

TANGELO
Minneola

KUMQUAT
Nagami

45 degrees with the trunk. Subsequent pruning need only remove deadwood, crowded branches and water sprouts (erect fast-growing shoots from the trunks of trees).

Citrus trees are sensitive to wet soil and overwatering, yet irrigation is necessary in many citrus-growing areas; to be sure the trees have adequate drainage, plant them on a slight mound with a 3- to 4-inch permanent earthen basin around the trees. When water is put in the basin, it should not rise to the level of the trunk. Watering in desert areas should be deep enough to wet the soil to a depth of 4 feet or more. Each watering should be very thorough or salts will build up in the soil and damage the roots; deep watering leaches the salts away. Maintain a 2-inch mulch of grass clippings, sawdust, wood chips or ground bark on the soil around citrus trees.

In Arizona and California the primary fertilizer needed is nitrogen. To supply this nitrogen, give young trees during their first two years a total of 10 ounces each year of ammonium sulfate in three equal applications in early spring, late spring and late summer. The amount of ammonium sulfate applied each year thereafter should be increased at the rate of 1 pound per year until trees reach maturity. In Florida, Louisiana and Texas a complete fertilizer specifically designed for citrus trees is recommended. During the first and second years, apply a total of 2 pounds of 8-8-8 fertilizer in three equal applications, one in early spring, one in late spring and one in early fall. After the first two years, increase the fertilizer to 3 pounds per year of the tree's age until maturity. Citrus trees sometimes suffer from deficiencies of zinc or iron. The symptom of a zinc shortage is mottled green-and-yellow leaves; to correct the lack, spray a solution of zinc oxide—½ pound for every 10 gallons of water—on the leaves when they are nearly fully open. Lack of iron in a form usable by plants causes iron chlorosis, which appears as pale green leaves with dark green veins. It can be corrected by spraying the leaves with a solution of 1 pound of chelated iron 138 to 10 gallons of water; it is advisable to add a teaspoonful of laundry detergent to help the solution spread. About 6 to 7 gallons of the solution are generally needed.

Because citrus fruits do not improve in flavor after they are picked, they should be allowed to ripen on the tree. Fortunately, most citrus fruits cling to the trees for months after they are ripe, and even though their flavor slowly deteriorates, the harvest season is spread over weeks or months. Citrus fruits are fully ripe when they have reached the color and size specified for their type. All fruits should be picked with shears (and gauntlet gloves if the trees have thorns) and should be cut off flush with the button, the point at which the fruits are attached to the stem.

CURRANT
Ribes

The currant is a seldom-grown fruit, partly because its berries are too tart to eat fresh; they do, however, make excellent jams. Like the gooseberry the currant spreads white-pine blister rust, a fungus disease that causes extensive damage to pine forests. Because of this, cultivation of currants is prohibited in parts of the country where white pines grow. Check with your county agricultural extension service to find out if this affects your area.

Currants are cold-climate fruit and grow in Zones 3-6. The bushes usually grow about 3 to 4 feet tall and spread to an equal distance. The red or white berries, borne in clusters in midsummer, are about ¼ inch in diameter. The more popular currants are the red ones; good varieties are Perfection, Red Lake and Wilder. A good white variety is White Imperial. Plants will bear crops for 20 to 30 years, and each usually yields 4 to 6 quarts of fruit annually.

HOW TO GROW. Currants grow best in full sun or light shade and do best in a moist, medium to heavy soil of pH 5.5 to 7.0 that has been supplemented with compost or manure. For berries within two or three years, buy one- or two-year-old plants and set them out in the fall. Plant them 5 feet apart and cut them back to 6 to 8 inches from the ground; no further pruning is necessary for the first three years. Each fall or winter thereafter cut back to ground level all canes over three years of age, and remove all but six to eight new canes. If a manure mulch is applied each fall, no additional fertilizer is needed; otherwise, scatter a handful of 5-10-5 fertilizer beneath each plant in the fall or early spring.

Harvest currants when they are firm and their color is developed but before they become fully ripe; they then have the highest pectin content and are best for jam making. To pick them, grasp the stems at the tops of the cluster, twist the clusters off, then strip the berries from the clusters. To propagate new plants, set 8- to 10-inch pieces of one-year-old stems directly into the ground in the fall, burying all but two buds on each cutting.

F

FIG
Ficus carica

The delicately sweet 1- to 3-inch fig is one of the oldest fruits known to man. The trees may grow 6 to 20 feet tall and spread to an equal or greater distance. The handsome deeply lobed leaves, often used as a motif in art, are borne on thick stubby twigs; the trunks and branches are covered with pearl-gray bark and can become attractively gnarled as the trees become old. Most figs are brown, purplish or pale shades of yellow or green when ripe. The fruit itself is unusual in that the edible part we call the fig is not the true fruit but a fleshy receptacle whose inner walls are lined with tiny seeds; these are the true fruit.

Figs grow normally only in Zones 7-10, but established trees that have not been heavily fertilized will survive temperatures as low as 15° without injury, and in cold areas some gardeners protect the branches from freezing by encasing the trees completely in straw-and-burlap "mummies." In the Southeast and along the Gulf Coast west to Texas, recommended varieties are Brown Turkey, also called Black Spanish and San Piero; Brunswick, also called Magnolia; Celeste, also called Blue Celeste; and Texas Everbearing. In California the best home-garden varieties are Brown Turkey; Mission; Brunswick; Genoa, also called White Genoa; and Kadota. In the Northwest a recommended variety is Lattarula, also called White Italian. All produce their edible parts—the so-called fruit—without pollination and so can be planted alone.

Fig trees produce two crops a year, one in early summer from buds on the previous season's growth and another in late summer on the current season's growth; occasionally a third crop is borne on the late summer's branches. Fig trees may bear for 50 years or more, and a 15- to 20-foot tree ordinarily yields at least 40 to 50 pounds annually.

HOW TO GROW. Figs are tolerant of many soil types but do especially well in moist but well-drained clay soil with a pH of 5.0 to 6.5. For fruit in the second year, buy trees 3 to 4 feet tall. Plant them in the winter or early spring, being careful that the roots do not dry out during the planting operation. Cut off all but three or four of the best-placed branches; these should be 6 to 12 inches apart, face in different directions and form angles greater than 45 degrees with the trunk. Cut them back to 6 to 8 inches. Mulch around newly planted trees with a 2- to 4-inch layer of

CURRANT
Red Lake

For climate zones and frost dates, see maps, pages 148-149.

wood chips or ground bark. The area under older trees can be planted with grass. Fig trees seldom need fertilizing, but if the leaves are pale or yellow-green, scatter a few handfuls of lawn fertilizer beneath them in winter or early spring. Young trees should be pruned during the dormant winter season only enough to train them to the desired shape; old trees seldom need pruning.

Figs are seldom bothered by insects or diseases in home gardens, but as birds are very fond of figs, it may be necessary to cover the trees with plastic netting to protect the ripening fruit.

Figs should be picked when the necks of the fruit shrivel so that the fruit hang straight down; if white sap appears when a fruit is picked, it has been picked too soon. Figs for drying should be allowed to fall from the tree, at which time they will be partly dehydrated; finish drying them by spreading them on trays in the sun.

To propagate fig trees, insert 12-inch twigs into the ground in the winter so that only one bud shows above the soil surface; in the spring a new tree will start to grow from each cutting.

G

GOOSEBERRY
Ribes

Gooseberries are seldom eaten fresh but are prized for jam and pie. However, like currants, they spread white-pine blister rust, and their cultivation is prohibited where white pine trees grow. Ask your agricultural agent if gooseberries can be legally grown where you live.

The gooseberries recommended here, all American varieties, can be grown throughout Zones 3-7. They grow on very cold-resistant, thorny 2- to 4-foot bushes. The ¾-inch berries ripen in midsummer and are usually greenish yellow, pink or red. The pink and red varieties are often sweet enough to eat fresh when they are fully ripe, but the greenish ones are quite sour. Good varieties with sweet berries, few thorns and large crops are Pixwell, greenish pink; Poorman, red; and Welcome, pink. Plants will bear berries for 25 years, and a mature plant ordinarily yields 5 to 10 quarts annually.

HOW TO GROW. Gooseberries grow in full sun or light shade in Zones 3-7, but do best in partial shade in the southern part of their growing area. They thrive in a moist highly organic, medium to heavy soil with a pH of 5.0 to 6.5. For berries within two years, buy one- or two-year-old plants. Plant them about 5 feet apart in the fall, and set them 2 to 3 inches deeper than they grew in the nursery. Cut the plants back to 6 to 8 inches above the ground. During the first winter cut the original stubs and all but the six strongest canes to ground level; during the second winter cut out all but the original six canes plus the three strongest canes of the current year. During the third winter prune out all but the three strongest canes of that year's growth plus three of the six canes of the first year's growth. Each winter thereafter cut out the three oldest canes and all but three of the current season's growth; thus each plant will produce fruit on one-, two- and three-year-old canes.

Unless a manure mulch is applied each fall, scatter a handful of 5-10-5 fertilizer beneath each bush in the fall or early spring. To pick gooseberries, put on long leather gloves and strip the fruit from the canes. New plants can be propagated by the method known as ground layering. Bend a cane so that it touches the ground about a foot in from the tip, bury a part of this section that contains a leaf joint, or node, setting it 3 inches deep and anchoring it with crossed sticks or a rock. Make sure three buds of the cane extend beyond the buried portion. During the sum-

FIG
Brown Turkey

mer roots will form on the buried stem and new growth will appear on the three exposed buds. Early the following spring before growth starts, free the new plant from its parent by cutting the cane near the new roots; dig it up and plant it elsewhere. Do not let the roots dry out.

GRAPE

American bunch grape, also called fox grape *(Vitis labrusca)*; European grape *(V. vinifera)*; muscadine grape *(V. rotundifolia)*

The grape is one of the oldest fruits known to man, and one variety or another can be grown in almost every part of the country. Grapes grow on vigorous vines that cling by tendrils to any available support. The 4- to 8-inch leaves are roughly heart-shaped and the ½- to 1-inch fruit, borne in clusters, ripen from midsummer until late fall according to the variety. Grapevines may live for 100 years or more, and a mature grapevine may bear as much as 10 to 20 or more pounds of fruit annually.

American bunch grapes and their hybrids, which are the result of crossing American and European varieties, are adaptable to wide differences in climate. Grapes of this type and also of the muscadine species are sometimes called slipskin because the flesh and the skin separate easily; the skin is usually not eaten.

The varieties of American grapes recommended below are suited to Zones 5-8 except where noted. All but Brighton pollinate themselves so that they can be planted singly. Early grapes, which ripen in late summer or early fall, are Brighton (Zones 5-9), a good variety of red grape; Buffalo, blue black, good for wine and virtually pest free; Himrod Seedless (Zones 5-9), yellow; Interlaken Seedless (Zones 5-9), amber; Ontario, light yellow, productive and virtually pest free. Midseason grapes, ripening in early fall, are Delaware, red, good for wine; Golden Muscat, green, excellent for jelly and juice; Niagara, gold, good for arbors; and Van Buren (Zones 4-8), purple, good for jelly and juice. Recommended late grapes, which ripen in late fall, are Alden, large sweet blue grapes good for wine; Catawba (Zones 5-9), red, keeps well and is good for wine; and Concord, blue black, excellent for jelly and juice.

European grapes make up over 90 per cent of the nation's commercial grape crop. Some varieties do well in Oregon and eastern Washington, but generally they are at their best in warm interior valleys of California and at low elevations in Arizona and New Mexico. All of the varieties listed here are dessert grapes; they are all self-pollinating, and unless otherwise noted, all grow in Zone 9. Recommended early-ripening grape varieties are Cardinal, dark red; and Pearl of Csaba, also called Csaba (Zones 6-9), amber white, nearly seedless. Early midseason grapes are Blackrose, a large black grape; Red Malaga, also called Molinera, large purple-pink fruit good for arbors; and Ribier (Zones 7-9), large sweet black fruit. Midseason grapes are Black Monukka, also called Monukka (Zones 6-9), a reddish black medium-sized seedless fruit well suited to the Northwest; Thompson Seedless, also called Sultanina, a golden seedless grape good for raisins and wine; and White Malaga, gold. Late midseason grapes are Muscat, also called Muscat of Alexandria, green, good for eating fresh and for raisins; Olivette Blanche, also called Lady Finger, a green oval; and Tokay, also called Flame Tokay, brilliant red and large, especially suited to the San Joaquin Valley. A late-ripening grape is Emperor, red.

Muscadine grape varieties can be grown throughout Zones 7-10. The vines grow very vigorously and the fruit are large with few in each cluster. Unlike most other grapes, many of the cultivated varieties of muscadine grapes bear

GOOSEBERRY
Pixwell

For climate zones and frost dates, see maps, pages 148-149.

only female flowers. They must have a self-pollinating variety planted close by, for pollination by the pollen from one of these varieties is necessary for the vines to bear fruit. One self-pollinating variety will provide enough pollen for a dozen female vines. The following muscadine grapes are all female except as noted. Recommended early-ripening types are Hunt, black; and Scuppernong, bronze green. Good varieties of midseason grapes are Burgaw, reddish black, self-pollinating; Dulcet, purple; Tarheel, black, self-pollinating; Thomas, wine red; Wallace, bronze yellow, self-pollinating. Late varieties are Creek, reddish purple; James, black; and Yuga, golden red.

HOW TO GROW. Grapes grow in a wide variety of soils but do best in sandy, highly organic soil of pH 5.5 to 7.5 that has been supplemented with compost or manure. Heavy clay soil usually induces many leaves but relatively few, inferior grapes. Plant two-year-old vines in very early spring as soon as the ground can be worked. Grapevines generally require about four or five years to reach full productivity, but often begin to bear a small amount of fruit the second or third season after planting. Space American and European vines about 8 feet apart both ways, muscadine vines about 20 feet apart in rows 15 feet apart. For easier harvesting of muscadine grapes, grow them on arbors tall enough to walk beneath. Other grapes may also be grown in this way, as well as on fences and trellises.

Grapes are borne from buds produced on canes that grew during the previous summer; vines must be pruned before the sap flows in early spring to reduce the number of buds so that the ones that remain will produce many good-sized bunches. Each pruning leaves some buds that will produce the current season's fruit, as well as other buds that will produce canes to bear the following season's crop. Immediately after planting, cut the plants back to a single cane with two strong buds. During the first summer train the young vine up a 5-foot stake set beside it and remove the weaker of the canes that grow from the two buds.

In the spring of the second season, grapevines require specialized pruning. Grow muscadine varieties on arbors about 7 feet tall. As each main stem reaches the top of the arbor, it will develop about eight canes, or arms, that spread across the top of the arbor. The arms send out lateral, or side, branches each year; prune these one-year-old laterals back to four to six buds each in early spring. They will bear the current season's crop.

Although American grapes can similarly be grown on arbors, they are best trained by the four-cane Kniffen system. In this system the vine is pruned to four canes, each of which is tied to a wire (drawings, page 52). The four-cane Kniffen system can also be used to train some of the European grape varieties, such as Black Monukka, Blackrose, Emperor, Olivette Blanche, Red Malaga, Thompson Seedless and White Malaga. Usually, however, European grapes are better when grown on a single wire strung between posts about 3 to 3½ feet from the ground. To follow this system, each year cut away all but four canes. Tie two to the wire and cut them back to 6 to 10 buds each; cut the other two back to two buds each. The tied canes will produce the current year's fruit, and the two-bud canes will produce canes for the following year's fruit.

A type of pruning that is especially suited to such European varieties as Black Monukka, Blackrose, Cardinal, Emperor, Muscat, Pearl of Csaba, Red Malaga, Ribier, Tokay and White Malaga, as well as most wine grapes, is called spur pruning. Each vine is grown against a single stake, and each year the growth of the current season is cut back to two or three buds (drawing, page 53). In parts of the Southwest where there is a deficiency of zinc in the

GRAPE
Left: Buffalo   *Right:* Golden Muscat

soil, daub each cut with a zinc-sulfate solution, 1 pound to a gallon of water, within two hours after pruning.

If the stems of the grapevines seem weak, feed plants in early spring with ¼ to ½ pound of ammonium sulfate scattered widely beneath each vine.

Except for muscadine varieties, whose fruit are shaken from the vines onto a cloth, grapes should be harvested by cutting off entire bunches with small hand shears. For best flavor in dessert grapes and for making grape juice or wine, allow the fruit to attain full ripe color on the vine. For jelly making, pick them when they are a bit underripe because they have a higher pectin content at that time and produce crystal-clear jelly. Fresh grapes stored dry in a refrigerator will keep for one to two months.

Grapes can be propagated in spring by the method known as ground layering. Bend a cane so that it touches the ground about a foot in from the tip; bury a part of this section that contains a leaf joint, or node, setting it 3 inches deep and anchoring it with crossed sticks or a rock. Make sure three buds of the cane extend beyond the buried portion. During the summer roots will begin to form on the piece of buried stem and new growth will appear on the three exposed buds. Early the following spring before growth starts, cut the new plant from its parent, dig it up and plant it elsewhere. Do not let the roots dry out. Alternatively, grapes can be propagated from the hardwood cuttings of healthy dormant stems.

GRAPEFRUIT See Citrus

K
KUMQUAT See Citrus

L
LEMON See Citrus
LIME See Citrus
LIMEQUAT See Citrus
LOGANBERRY See Blackberry

N
NECTARINE See Peach

O
ORANGE See Citrus

P
PEACH and NECTARINE
Prunus persica, also called *Amygdalus persica*

Peaches and nectarines are essentially the same fruit, their primary difference being that peaches are fuzzy and nectarines smooth-skinned. The smooth-skin characteristic that distinguishes nectarines is a minor genetic variation, like red hair among people; it is even possible that a peach tree may suddenly produce a branch that bears nectarines, and vice versa.

Whatever their name or the condition of their skin, these tree fruits are the ones with which the average home gardener is most apt to succeed, for they are fairly easy to grow over much of the country. Peach and nectarine trees are available as standard trees, which may grow 12 to 15 feet tall if unpruned, and as two kinds of dwarfs—one that seldom grows more than 3 or 4 feet tall, and another that grows to 6 to 7 feet. For most home gardeners the standard tree is the best one to buy because it is the least expensive, its size can be easily limited to 7 to 10 feet by conscientious pruning and it bears the greatest amount of fruit for the space it occupies. The flowers appear before the leaves come out in spring, and range in color from pale

GRAPE
Left: Olivette Blanche   *Right:* Ribier

For climate zones and frost dates, see maps, pages 148-149.

PEACH
Reliance

pink to dark red. The fruit generally ripens in midsummer.

Peaches and nectarines were originally moderate-climate fruits but new varieties can be grown in most parts of the country except on the plains and mountains west of northern Illinois and north of Colorado and in the warmest parts of Florida and the Southwest. One peach variety, Reliance, will bear a full crop of fruit after winter temperatures that reach as low as 25° below zero.

Most recommended varieties have red-and-yellow skins; the colors referred to here pertain to the flesh of the fruits. Peaches are usually classified as clingstones, with flesh that clings to the pit; freestones, with flesh that separates easily from the pit; and semiclingstones, with flesh that clings to the pit before fully ripening, then is generally free. Peaches recommended for the lower South and Southwest (Zones 7-9) are Earligold, yellow, semiclingstone; June Gold, yellow, clingstone; May Gold, yellow, clingstone; Sam Houston, yellow, freestone; and Southland, yellow, freestone. Two peach varieties for Zone 9 are Early Amber, yellow, clingstone; and Flordasun, yellow, semiclingstone. Peach varieties suited to Zones 6-8 are Halehaven, Redhaven, Redskin, Richhaven and Sunhaven; all have yellow flesh and easily removed pits. For cold regions (Zones 5-8) the recommended varieties are Harbelle, Reliance, Stark Frostking and Sunapee, all with yellow flesh and easily removed pits; and Stark Surecrop, white, clingstone. The most important nectarine varieties, which grow in Zones 6-8, are Cavalier and Freedom, yellow, freestones; Nectarheart, white, freestone; and Redbud and Red Chief, white, semiclingstones. A variety of nectarine planted in Zone 9 in Florida is Sunred, yellow, semiclingstone.

Trees of at least two varieties should be planted, for pollination of one by the pollen from the other is usually necessary if the trees are to bear fruit. Trees may live 10 to 20 years; they may survive much longer, but they decline in vigor as they become older. When only four or five years old, one tree may produce 2 to 3 bushels of fruit annually; by the time it reaches 8 to 10 years of age, it produces 4 to 6 bushels annually.

HOW TO GROW. Peach and nectarine trees grow best in soil with a pH of 6.0 to 7.0. For fruit within two or three years, buy two-year-old 3- to 5-foot trees. Plant them in early spring on an elevated or sloping site to prevent flower buds from being killed by spring frosts. Cut the trees back to 2½ to 3 feet above the ground at the time of planting. Cut away all but the three strongest remaining side branches, then cut them back to 2- to 4-inch stubs. The stubs should be 6 to 12 inches apart, face in different directions and form angles of 45 degrees or more with the trunk. In subsequent years prune in early spring to remove low-hanging, crowded or crossing branches. Cut back overly long branches to control tree height.

During the first spring, scatter 1 pound of 10-10-10 fertilizer in a wide circle around each tree, and repeat once about six weeks later. Use a water-soluble fertilizer, not the urea-type lawn fertilizer, which releases nitrogen so slowly that it stimulates late-season growth of young wood easily damaged by winter cold. Increase the amounts gradually in subsequent years, making a single application early in the spring. Mature trees may use up to 5 pounds of 10-10-10 fertilizer annually, but the best indicator of fertilizer need is the leaf color. If it is dark green, no feeding is needed; if pale or yellow-green, use the fertilizer.

Because peaches and nectarines both produce more fruit than they can support, it is necessary not only to thin out the branches by pruning as noted above but to thin out the fruit as well. Otherwise the crop may break the branches or be undersized and have inferior flavor and color. Each peach or nectarine should be balanced by about 35 leaves, a proportion achieved by thinning fruit to hang 6 to 8 inches apart on the branches.

Peaches and nectarines generally ripen in midsummer and should be picked when they can be separated from the twigs with very little effort. Handle them carefully to prevent bruising. Home-frozen peaches that have been allowed to ripen fully are nearly equal in quality to those freshly picked from the trees.

PEAR
Pyrus

Pears are a favorite of home gardeners because they are attractive, suited to small spaces and indeed grow best on a lawn. Full-sized trees can be kept pruned to 15 to 18 feet and dwarfs grow only 8 to 10 feet. The leaves are noted for their shiny, glossy surface, and the blossoms are white.

Nearly all pears grown for eating raw are descended from *P. communis,* a European species. Hybrids between *P. communis* and the sand pear *P. serotina,* also called *P. pyrifolia,* an Oriental species, are not as flavorful as are common pears, and they have the defect of producing gritty stone cells in the flesh of the fruit. They are quite good for cooking, however, and these varieties will tolerate warmer winters than will common pears, doing reasonably well as far south as Zone 9. Many hybrid pears are very resistant to a bacterial infection called fire blight that affects most common pears.

Unless noted otherwise, all the varieties listed here are common pears that grow well from coast to coast in Zones 5-7. A recommended early-ripening pear is the yellow Clapp's Favorite. Midseason pears are Bartlett, yellow, the most popular of all pears; Max-Red Bartlett, similar to Bartlett but red-skinned; Bosc, also called Beurre Bosc and Golden Russet, yellow; Duchess, also called Duchess d'Angoulême, yellow; Magness, blight resistant, yellow; Orient (Zones 5-9), a russet blight-resistant hybrid pear of special usefulness in the South; Seckel (Zones 5-8), an old-fashioned, small, golden-brown, blight-resistant pear with a honeylike flavor; Starking Delicious (Zones 5-8), a large yellow blight-resistant variety; and Tyson (Zones 5-8), a small sweet yellow pear. Recommended late-ripening varieties are Anjou, also called Beurre d'Anjou and d'Anjou, blight resistant, greenish brown; Comice, also called Doyenne du Comice and Royal Riviera (Zones 5-8), yellow, especially suited to the West Coast; Kieffer (Zones 4-9), a large yellow blight-resistant hybrid widely used for canning; LeConte (Zones 5-9), a blight-resistant greenish yellow hybrid especially suited to the South; Moonglow (Zones 5-8), a yellow blight-resistant hybrid; and Winter Nelis (Zones 5-8 on the West Coast), a small yellow pear with a superb flavor.

Trees of at least two different varieties should be planted, for pollination of one variety by pollen from another is usually necessary to ensure that the trees bear fruit. Most pear varieties listed except Bartlett and Seckel will pollinate each other: Duchess is self-pollinating and can be planted alone; Magness must be planted with Duchess or with two other varieties that will pollinate each other. Pear trees are extremely long-lived, often bearing fruit for more than 100 years. A mature standard tree produces 5 to 10 bushels of fruit annually and a dwarf ½ to 1½ bushels.

HOW TO GROW. Pears grow best in a heavy, moist but well-drained soil with a pH of 6.0 to 7.0. Plant one- or two-year-old trees. Standard trees of that age are 4 to 6 feet tall and will bear fruit within four or five years; dwarf trees of similar age are 2½ to 4 feet tall and will often bear fruit the second or third year. Cut young standard

PEAR
Left: Bosc   *Center:* Bartlett   *Right:* Seckel

For climate zones and frost dates, see maps, pages 148-149.

PERSIMMON
Hachiya

trees back to 3 feet at the time of planting and shorten the branches on both standard and dwarf varieties by one half. Train the trees as they grow so that they will have four to eight well-distributed branches arising from the trunk, spaced 6 to 12 inches apart; the angle between each branch and the trunk should equal 45 degrees or more. Prune trees as little as possible after the skeleton branches have been established. It is important that pear trees carry only as much fruit as they can support. Part of the problem of overproduction will be taken care of naturally, as some of the young fruit will drop about six weeks after the flowers bloom. Thin out the remaining fruit, saving the best of each cluster, so that pears are about 6 to 8 inches apart.

In many gardens pear trees do not need fertilizer; if the foliage, however, is pale or yellowish green, use 1 pound of 10-10-10 fertilizer per year of the tree's age, scattering it under the spread of the branches. Pears should not be mulched but should be grown on a lawn area where they must compete for food and moisture with grass. The chief reason for deliberately undernourishing pear trees is to forestall damage by a bacterial infection called fire blight that attacks trees that are growing too luxuriantly. This disease may be carried from tree to tree by bees at the time of blossoming; within a few weeks shoots develop black leaves and begin to die back. If fire blight is detected, cut off the diseased portions, making each cut at least 12 inches beyond the apparent injury. Wipe saws and clippers with alcohol after each cut.

Pears, unlike most other fruits, cannot be allowed to ripen on the tree; tree-ripened pears develop brown centers, and the flesh becomes soft and off-flavored. Pears are ready to harvest when their stems begin to swell at the point of attachment to the twig and when the dark green color shows signs of yellowing. Pick pears by giving them an upward twist; if the twigs break rather than separate cleanly from the fruit stems, the pears are not ripe enough to pick. Store the fruit in a cool place for one to two weeks before bringing them into a warm, dark room to ripen.

Several pear varieties can be grown on the same tree by grafting different varieties to the branches.

PERSIMMON
Kaki persimmon, also called Japanese persimmon *(Diospyros kaki);* common persimmon *(D. virginiana)*

Persimmons are delectable fall fruit for home gardeners in Zones 6-10. Although unpleasantly astringent when hard and immature, a fully ripe persimmon is soft and sweet, with jellylike flesh. The kaki species is especially tasty. These persimmons are heart-, plum- or tomato-shaped, 2 to 4 inches in diameter and golden red, yellow or orange in color; the flesh may be yellow or brown. The trees grow 20 to 30 feet tall with an equal spread; they are especially beautiful in autumn when their leaves turn brilliant shades of yellow and red. Good varieties are Chocolate; Eureka; Fuyu, also called Gaki or Fuyugaki; Hachiya; Tamopan; and Tane-Nashi. Japanese persimmon trees are usually self-pollinating and can be planted alone.

The common persimmon grows wild in much of the southern and eastern half of the United States and will grow in Zones 5-10. Trees grow 30 to 60 feet tall with a spread of 20 to 30 feet. This species bears 1- to 2-inch yellow, orange or purplish fruit in the fall. Good varieties are Early Golden, Garrettson and Killen. Plant trees of at least two varieties to ensure that the trees bear fruit.

Persimmon trees may live for 50 years or more, and a mature tree yields 75 to 100 pounds of fruit annually.

HOW TO GROW. Persimmon trees grow best in soil with a pH of 6.0 to 7.0. For fruit within three years, buy one- or two-year-old trees. Set out container-grown trees at any time, but plant bare-rooted trees in early spring as soon as the ground can be worked. Cut off all but three to five well-placed branches, spaced 6 to 12 inches apart and forming angles greater than 45 degrees with the trunk. Cut these branches (on a bare-rooted tree only) to 6 to 8 inches. Beginning the second spring, apply 1 pound of 10-10-10 fertilizer for each year the tree has been planted, up to a maximum of 5 pounds per tree. Thin out the fruit until the persimmons are 6 to 7 inches apart.

Insects and diseases rarely bother persimmons. The fruit should be picked when soft by snipping them from the branches with shears; a small stem should be attached to each fruit. The fruit will continue to ripen off the tree.

PLUM

European plum *(Prunus domestica),* damson plum *(P. domestica insititia),* Japanese plum *(P. salicina)*

There are plums that will grow in every state in the United States as well as in most of southern Canada. Nearly all are good for eating fresh and for canning as well as for making preserves; some varieties can also be dried as prunes. Nearly all prunes are produced on the West Coast, where the warm temperatures and low humidity permit the fruit to dry readily. Most standard-sized plum trees may reach a height and spread of about 15 to 20 feet; dwarf trees grow only 8 to 10 feet tall.

The following varieties of European plums are recommended; all will grow in Zones 5-7, and Burbank Grand Prize prune will also grow in Zone 8. All ripen in midsummer: Blufre prune, blue; Burbank Grand Prize prune, purple; Green Gage, also called Reine Claude, greenish yellow; Stanley prune, purple; and Yellow Egg, bright yellow. Green Gage, Stanley and Yellow Egg are self-pollinating and will bear fruit if planted alone. Plant Blufre and Burbank Grand Prize (or other European varieties) together, for pollination of one variety by the pollen from another is usually necessary for the trees to bear fruit.

Damson plums are generally too tart to eat fresh unless very ripe, but make excellent preserves. French Damson and Shropshire are two varieties particularly recommended. Both have small purple-skinned, green-fleshed fruit that ripen in late summer. They grow in Zones 5-7, and either can be planted alone.

Japanese plums have juicy fruit that are good fresh or cooked. Except for Redheart, which grows in Zones 6-9, the varieties listed grow in Zones 5-9: Ozark Premier, bright red, midsummer; Redheart, red, midsummer; Santa Rosa, red, early midsummer; Shiro, yellow, early midsummer; and Starking Delicious, red, midsummer. Plant at least two of these varieties together, for pollination of one Japanese plum tree by the pollen of another is necessary for the trees to bear fruit.

Hybrid varieties created by crossing Japanese and American plums are particularly useful to northern gardeners, for these types grow even in Zone 4 as well as in warmer regions as far south as Zone 8. Recommended varieties are Superior, red, midsummer; Tecumseh, red, early summer; Underwood, red, early summer; Waneta, purple, midsummer; and Kaga, red, early summer. The Kaga variety should be included in every planting because it is the best pollinator for the other varieties listed.

Plum trees will bear fruit for about 25 to 30 years; a mature standard tree yields 1 to 2½ bushels annually, and a dwarf tree yields ½ to 1 bushel annually.

HOW TO GROW. Plums need soil with a pH of 6.0 to 8.0. Buy standard trees 3 to 6 feet tall, and dwarf trees 3 to 4 feet tall. Japanese and Japanese-American hybrids begin

PLUM
Clockwise from top left: Green Gage, French Damson, Santa Rosa, Stanley

For climate zones and frost dates, see maps, pages 148-149.

to bear two to four years after planting; European and damson plums begin to bear three to five years after planting. In Zones 4 and 5, plant plum trees in early spring as soon as the ground can be worked; in Zones 6-9, they should be planted during the fall or winter. Cut off all but three or four of the strongest branches that are spaced 6 to 12 inches apart; make sure they spread in different directions and form angles greater than 45 degrees with the trunk. Cut them back to 6- to 8-inch stubs. Each year thereafter prune the dormant trees only to thin overcrowded or crossing branches or to remove deadwood or the erect fast-growing shoots called water sprouts that appear along the trunk or branches. When pruning, be sure to remove any stems that have black swellings, called black knot. Make cuts at least 6 inches beyond any apparent injury.

Fertilize each plum tree with 1 pound of ammonium sulfate in the spring of the first year; scatter the fertilizer in a 4-foot circle around the tree after new growth begins. Fertilize each spring thereafter, and gradually increase the amount so that a mature standard tree receives 2 to 3 pounds and a dwarf tree receives 1½ to 2 pounds annually. European, Japanese and hybrid plum trees often bear larger crops than they can support. In early summer, when the fruit are about one third grown, thin the fruit so that the plums are 2 to 3 inches apart. Do not thin damsons.

For cooking purposes pick plums when they become covered with a waxy white coating called bloom and are firm but springy to the touch. For eating fresh or for drying, pick them when they become soft and fully ripe and are easily twisted off.

R

RASPBERRY
Rubus

Although raspberries are easy to grow, they are too fragile and perishable to be widely marketed, and so are especially valuable in home gardens. They grow on erect shrubs that bear loose clusters of white flowers in early summer; the flowers are followed immediately by the fruit. Black raspberries, often called blackcaps, and purple raspberries, which are hybrids between red and black varieties, have quite a different flavor from that of red raspberries. Yellow raspberries are mutations of red varieties and are like them except for color.

There are two categories of red and yellow raspberries, the more important being the common raspberry, which ripens in early to midsummer. The other type is called fall or everbearing; although lacking the subtle flavor of the summer varieties, it produces both an early-summer crop on the previous season's growth and a fall crop on the current season's growth. Excellent varieties of summer-fruiting red raspberries are Canby, Hilton, Latham, Taylor and Willamette; of these, Canby and Willamette grow in Zones 5-8, while the others grow in Zones 4-8. Recommended varieties of everbearing red raspberries include Heritage, Fallred, Durham, Indian Summer and September; all grow in Zones 4-7. Two summer-fruiting yellow raspberries are Amber and Golden West; an everbearing yellow raspberry is Fall Gold. Among the recommended varieties of black raspberries, all of which grow in Zones 5-8, are Allen, Black Hawk, Bristol, Cumberland and Dundee. Outstanding purple raspberries are Clyde and Sodus.

A row or hill of raspberries will ordinarily produce good crops of fruit for 10 years or more, and the plants usually yield one quart of fruit annually to each foot of row.

HOW TO GROW. Raspberries grow best in well-drained soil of pH 5.5 to 7.0 that has been supplemented with compost or manure. Buy one-year-old certified disease- and virus-free plants. Raspberries should never be planted where eggplants, peppers, potatoes or tomatoes have grown within three years, because they are susceptible to soil-borne diseases associated with such plants. Spring planting is recommended. If plants are to be grown in rows, set red and yellow varieties about 2 to 2½ feet apart in rows about 7 to 8 feet apart; set black and purple varieties 3 feet apart in rows 8 feet apart. If raspberries are to be grown as clumps, or hills, space the red and yellow varieties 5 to 6 feet apart and the black and purple ones 6 to 7 feet apart. After planting, cut the canes to 2 inches above the ground; leave the stubs to mark the rows until new sprouts appear from below the ground, then remove them.

Let the young canes of summer-fruiting red and yellow raspberries grow undisturbed until their second spring; when the buds begin to show green tips, remove all but two or three healthy canes per foot of row *(drawings, page 50)*. Hill plants should have all but six to eight canes removed. Cut the tops of the canes that will bear fruit to a height of 3 to 3½ feet.

While everbearing red and yellow raspberries will bear two crops a year, such production puts a great strain on the plants. It is best to cut the canes of everbearers to the ground late in the fall instead of in the spring or early summer, allowing them to make unrestricted growth during the summer so as to produce a heavy fall crop.

To prune black and purple raspberries, snip off the tips of the new canes in midsummer when they are about 3 feet tall. This will cause them to send out laterals, or side branches, that may more than double fruit production the following year. In the spring remove all but three to six strong canes and cut back the laterals to about 8 to 10 inches. Each of the buds on the laterals will bear several clusters of berries. Cut away the fruit-producing canes as soon as they finish bearing.

Feed raspberry plants in early spring by scattering 10-10-10 fertilizer around them at the rate of 1 pound per 10 feet of row. Birds are very fond of raspberries, but the fruit can be protected with plastic netting. Raspberries are ready for harvest when the berries separate easily from the stems. To pick them, gently pull each ripe berry between the thumb and forefinger; it will drop into your cupped hand. Handle the berries carefully and do not pile them into a deep container, for they crush easily. Do not wash or wet them because the water dilutes their flavor.

Red and yellow raspberry plants propagate themselves, spreading by underground suckers. If new plants are not wanted, pull out the suckers rather than cut them off, which only causes more suckers to grow. To get new black and purple raspberry plants, use the method called tip layering: cover the tips of the arching stems in the late summer with a shovelful of soil—a new plant will start to root at that spot in the spring. Cut off the tip, dig up the rooted plant (which may or may not yet show growth aboveground) and replant it elsewhere.

S

STRAWBERRY
Fragaria

Strawberries, the most popular fruit crop among home gardeners, are easy to grow in all parts of the United States and southern Canada. The plants grow 6 to 8 inches tall in a thick central crown from which emerge dark green three-leaflet leaves and fruit-producing stems; each plant will spread about 12 inches across, but runners may extend several feet beyond.

Most varieties of strawberries are especially adapted to a local climate; thus it is important to choose the ones that

RASPBERRY
Top: Latham   *Bottom:* Bristol

For climate zones and frost dates, see maps, pages 148-149.

STRAWBERRY
Ozark Beauty

will do best in your area. Strawberries are available in two types: June bearers and everbearers. The so-called June bearers, which include most varieties, produce one crop of fruit per year over a period of about two and one half weeks; they ripen in early to midsummer except in the cool coastal belt of California, where they produce fruit from April until November, and in Florida, where they ripen in midwinter. Although spring-planted June bearers produce blossoms their first summer, these are pinched off before they can set fruit, thus forcing each plant to use its energy to develop large amounts of fruit the following season. The plants will fruit well their third season, but lose their vigor thereafter and should be replaced.

The second type of strawberry, called everbearer, produces an early-summer crop and a fall crop, as well as some berries intermittently during the summer. When everbearers are planted in spring, the early-summer blossoms are removed that year only to prevent the plants from fruiting until fall; thereafter they will produce spring and fall crops each year until they lose vigor, usually after three years. Fall-planted everbearers begin to bear fruit in the winter or spring after planting; they too should be replaced after three years. In hot climates both types bear only one season and must then be replaced because they are debilitated by the heat.

Following are varieties recommended for home gardeners because of their particularly good flavor. They are listed by region in their order of ripening. In New York, New England and eastern Canada, good June bearers are Fairfax, Catskill, Surecrop and Sparkle; an everbearer is Ozark Beauty. Along the middle East Coast, recommended June bearers are Midland, Catskill, Surecrop and Vesper; an everbearer is Ozark Beauty. In the Southeast and along the Gulf Coast, June bearers are Suwannee, Pocahontas and Tennessee Beauty. In Florida, recommended June bearers are Florida 90 and Missionary. In the South Central region, June bearers are Earlibelle, Pocahontas and Tennessee Beauty; an everbearer is Ozark Beauty. In the North Central states and south-central Canada, June bearers are Midland, Surecrop and Sparkle; an everbearer is Superfection. In the Plains and Mountain states, June bearers are Cyclone, Dunlap and Trumpeter; an everbearer is Ogallala. In California and the Southwest, June bearers are Shasta, Tioga and Goldsmith; an everbearer is Red Rich. In the Pacific Northwest, June bearers are Marshall, Northwest and Siletz; an everbearer is Red Rich.

A single strawberry plant will provide about 1 pint of fruit per season.

HOW TO GROW. Strawberries do best in a highly organic soil of pH 5.5 to 6.5 that has been supplemented with compost or manure. Do not plant them where any tomatoes, potatoes, okra, melons, eggplants, cotton or raspberries have grown within three years, since they may pick up soilborne diseases remaining from such crops, and do not plant where grass has grown within the past year, since strawberries are harmed by grubs that may be present among grass roots. Work a 2- to 4-inch layer of well-rotted manure or compost into the top 8 inches of soil a few months before planting. Alternatively, apply a 2- to 4-inch layer of peat moss plus 1¼ pounds of 10-10-10 fertilizer per 100 square feet and dig the soil over thoroughly.

When setting out a new strawberry bed, buy only certified disease- and virus-free plants; do not replant your own, as they may have contracted a root disease. Plant strawberries in early spring in most of the country, but in Florida, along the Gulf Coast, in the Southwest and in coastal areas of California, plant them in fall. Depth of planting is critical (drawings, page 45); the object is to set each plant so that about one half its crown is buried and one half is above the soil. Immediately after planting, feed each plant with 1 pint of liquid fertilizer diluted to half the strength recommended on the label.

The so-called hill system of planting strawberries, practical for both the June-bearing and everbearing varieties, results in large berries but calls for considerable work. Space plants 12 to 18 inches apart in both directions, making beds two rows wide; the beds should be raised about 6 inches above the garden level to afford good drainage. Cut off all runners wherever they appear, so that all energy will be channeled to fruit production.

In cool areas where spring planting is the rule, an easier method of growing June-bearing varieties is the "spaced-matted" system. Set plants about 2 feet apart with rows 4 feet apart. Each plant will send out numerous runners that take root to form new plants. Allow no more than six new plants to develop, spaced about 6 to 8 inches apart. Cut off all surplus runners throughout the first season. The following season the plants will produce their best crop of fruit and will also make many new runners. Cut these new runners off only where they extend into the space between rows. In the spring of the following year start a new bed in a different spot, and plow the old bed under after the berries are harvested.

Use a rake or tined cultivator to stir the top inch of soil regularly to keep weeds away. June-bearing varieties planted in spring should have all their flower buds picked off before they open the first season; they will blossom and bear fruit the following summer. With spring-planted everbearers, pick off blossoms the first year until late summer, then allow them to produce a fall crop of fruit. Thereafter, let blossoms mature for a spring and fall crop each year. Everbearers planted in fall begin to bear fruit a few months after planting; do not remove their buds.

In the late summer, scatter 10-10-10 fertilizer around the plants at the rate of 1¼ pounds per 100 square feet of bed area. Keep fertilizer off the leaves, for it may burn them. Scratch it into the soil, or if the plants are mulched, spread it on top and water it in.

In the North, mulch the strawberry bed with straw or salt-marsh hay to a depth of 3 to 4 inches in fall when average night temperatures fall below 20°. Pull most of the mulch into the space between plants in spring when the new growth begins, leaving some around the plants to keep the fruit from resting on the soil. In the South, mulching is done mostly to keep fruit clean. In the fall, shortly after planting, apply a 1- to 2-inch layer of cottonseed hulls, peanut shells, sawdust, bagasse (sugar-cane fiber) or pine needles to cover the soil, not the plants.

To protect the berries from birds, which may eat them as soon as they begin to turn red, erect a low tent of plastic netting over a frame of stakes.

To pick fully ripe average-sized berries without bruising them, slip your index and second fingers behind a berry with its stem between your fingers, twist the stem a bit and pull with a sharp jerk; the stem will snap off about ½ inch from the berry. If the berries are large and thick-stemmed, cradle each one in your hand and pinch off the stem between your thumbnail and index finger.

T
TANGELO See Citrus
TANGERINE See Citrus
TANGOR See Citrus

Y
YOUNGBERRY See Blackberry

STRAWBERRY
Catskill

133

For climate zones and frost dates, see maps, pages 148-149.

Nuts

A

ALMOND

Prunus amygdalus, also called *P. communis* and *P. dulcis dulcis*

The almond, a relative of the peach, grows 15 to 30 feet tall with an equal spread. Like the peach, it has 3- to 6-inch leaves and bears pale pink or white blossoms on short spurs; even the fuzzy young fruit look like immature peaches. Instead of becoming plump with flesh, however, the fruit develop hard green husks *(center in the drawing)* that break open in the fall to reveal the inner brown shells *(below and right of center)* when the 1½-inch-long oval nuts *(below and left of center)* are ripe.

On the West Coast the principal varieties are Nonpareil, Ne Plus Ultra, Davey and Texas, also called Mission. Elsewhere the most common variety available is Hall, also called Pioneer and Ridenhower. Except where Hall is used, trees of at least two different varieties should be planted, for pollination of one variety by pollen from another is usually necessary if the trees are to bear fruit. Recent research shows that Hall is apparently self-pollinating, but it will set more fruit if peach trees are planted nearby. An almond tree, which can live and bear nuts for 50 years or more, will produce 25 to 40 pounds of nuts annually.

HOW TO GROW. The almond will grow in Zones 5-9 in any well-drained soil with a pH of 6.0 to 7.0. For nuts within three or four years, buy 2- to 4-foot trees and set them out in early spring. Because almonds bloom very early in the spring, their blossoms are sometimes killed by frost; to lessen this danger, plant the trees on high, sloping ground to allow cold air to flow away to lower elevations.

Cut off all but the three or four strongest, best-placed branches to form the skeleton of the tree; they should be 6 to 12 inches apart, spread in different directions and form angles greater than 45 degrees with the trunk. Prune these branches to a length of 6 to 8 inches; in subsequent years prune only to remove deadwood, crowded branches and fast-growing water sprouts along the trunk.

Almonds seldom need feeding, but if a mature tree gains less than 8 inches in height each year, scatter 1 to 1½ pounds of 10-10-10 fertilizer around it in the spring.

To harvest the nuts in fall, knock or shake them off the trees. Moisten the outer husks if they do not open easily, and crack open the softer inner shells. Dry the kernels in light shade and store them in airtight containers in a cool place. If frozen, the nuts will last several years.

B

BUTTERNUT See Walnut

C

CHESTNUT

Chinese chestnut *(Castanea mollissima)*

The Chinese chestnut is a fast-growing tree that eventually becomes 30 to 50 feet tall with an equal spread. It not only produces great quantities of delicious nuts but also has handsome glossy foliage that provides dense shade. Its slender, shiny leaves are 4 to 7 inches long and have toothed edges. They are reddish when they first unfold and turn shades of yellow and bronze in the fall. Young leaves are instantly identifiable by the dense, soft hairs on their surface. In early summer the trees bear masses of tiny fragrant white flowers in 8- to 10-inch upright clusters set among the leaves. The flowers are followed by 2- to 3-inch prickly seed husks that open in the fall, each releasing two or three shiny brown 1-inch nuts. Recommended va-

ALMOND
Nonpareil

rieties are Abundance, rich brown nuts; Crane, dark red, long-lasting nuts; Kuling, dark brown nuts; Meiling, tan nuts; and Nanking, dark tan nuts. All produce large nuts and, though not widely available, are worth seeking out.

Trees of at least two different varieties should be planted, for pollination of one variety by pollen from another is usually necessary if the trees are to bear fruit. Chinese chestnuts may live 50 years or more; a 10-year-old tree may produce 75 to 100 pounds of nuts annually.

HOW TO GROW. Chinese chestnuts grow in Zones 4-8 and do best in gravelly soil with a pH of 5.5 to 6.5. For nuts within two years, buy trees 4 to 5 feet tall. Trees should be set out in the spring. Cut off all but three or four of the strongest, best-placed branches to form the skeleton of the tree; they should be spaced 6 to 12 inches apart, spread in different directions and form angles greater than 45 degrees with the trunk. Prune these branches to a length of 6 to 8 inches. In the subsequent two or three years, remove any other branches that compete with these main ones; thereafter, pruning is usually unnecessary. Chinese chestnuts do best if weeds and grass are kept away from the trunks for the first three or four years.

Early in the second spring after planting, scatter 1 pound of 10-10-10 fertilizer over the area covered by the tree's branches. Each year increase the amount of fertilizer by 1 pound; large trees should receive 5 to 6 pounds per inch of diameter annually.

Gather nuts in fall, picking them up from the ground every day or two to prevent them from getting moldy. To test whether they are sound and full of kernel, put them in a pail of water. Save only those that sink to the bottom; after drying them, store in plastic bags in a refrigerator. The bags should be left partially open to allow moisture to evaporate from the nuts until they are dry. To freeze chestnuts, shell them and put the kernels in airtight containers; they will keep frozen for a year or longer.

F

FILBERT See Hazelnut

H

HAZELNUT (FILBERT)
Corylus americana, C. avellana

The hazelnut bears sweet-flavored nuts on easily maintained, relatively small trees that are 5 to 20 feet tall. The roundish tan nuts, which range about ½ to ¾ inch in diameter, ripen in the early fall.

The native American hazelnut *C. americana* is more resistant to winter cold than the European hazelnut *C. avellana*, but the European species produces the larger nuts. Many varieties are hybrids between these species and combine much of the hardiness of one species with the large nuts of the other. Varieties such as Bixby, Buchanan, Reed, Rush and Winkler grow throughout Zones 4-8 and are recommended for the area east of the Rocky Mountains. Along the West Coast recommended varieties are Barcelona, Brixnut, Daviana, Duchilly and Royal, which grow in Zones 5-8. Trees of at least two varieties should be planted, for pollination of one variety by pollen from another is usually necessary if the trees are to bear fruit; set them within 20 to 25 feet of one another. Hazelnuts live about 20 years, and when mature each tree will produce about 4 to 6 quarts of nuts annually.

HOW TO GROW. A moist well-drained soil with a pH of 6.0 to 6.5 is ideal. For nuts within three years, buy trees 4 to 5 feet tall. Since the flowers, which open very early in the spring, are susceptible to injury by cold and wind, plant the trees where they will be protected yet will not begin

CHINESE CHESTNUT
Abundance

HAZELNUT
Barcelona

For climate zones and frost dates, see maps, pages 148-149.

growth too early, such as on the northern slope of a hill.

In Zones 4 and 5, plant hazelnuts in very early spring as soon as the ground can be worked; in Zones 6-8, plant them in the fall or winter.

After new growth is well advanced during the first spring, scatter 1 pound of 10-10-10 fertilizer in a wide circle around each tree. In subsequent years gradually increase the amount by 1 pound; trees that are 15 to 20 feet tall should receive about 5 pounds of fertilizer annually.

If left alone to grow untended, hazelnut trees soon become thickets of slender stems because they send up great numbers of suckers, fast-growing stems that rise from below ground. If bush-type plants are desired, allow five or six stems to develop and remove the rest. Otherwise, suckers should be removed as soon as they appear, but they can be used for propagation (below). Most hazelnuts are grown as single-trunked trees, and this form is particularly advisable for the varieties grown on the West Coast. For a tree shape, cut off all but three or four main branches; these should be 6 to 12 inches apart, spread in different directions and form angles greater than 45 degrees with the trunk. Cut the branches back to 6 to 8 inches.

To prevent blue jays from eating the nuts before they are ripe, drape the trees with plastic netting. Once the ripe nuts have fallen to the ground, gather them up every day or two so that they will not become moldy or be harvested by squirrels. Test to see whether they are sound and full of kernel by putting them in a pail of water. Save only those that sink to the bottom; after drying them, store in a cool place. If shelled and then stored in airtight containers in a freezer, hazelnuts will keep for nearly a year.

Hazelnuts are propagated by hilling soil around suckers in the summer months. They will develop roots and be ready for digging up and planting elsewhere in spring.

P

PECAN
Carya illinoensis, also called *C. pecan*

Pecans are among the most flavorful and widely cultivated of all nuts. The trees are the largest members of the hickory family, growing 100 feet or more with massive trunks and stout, spreading branches. They grow wild from Illinois to Texas and Maryland to Florida, and have been widely planted elsewhere, especially in the Southwest and on the West Coast. Wild trees vary greatly in the size of the nuts they bear; they are sturdy, however, and their seedlings are used as the stock on which large-nut varieties are grafted. The dark brown nuts are covered with thin husks that break when ripe in early fall.

There are two general classes of cultivated pecan varieties: the southern ones, suited for Zones 7-10, which require about 250 frost-free days each year, and the northern ones, suited to Zones 5-7, which require 180 to 200 frost-free days each year. The chief southern varieties, which bear thin-shelled nuts about 2 inches long, are Desirable, Mahan, Moneymaker, Schley, Stuart and Success. Northern varieties bear nuts 1 to 1½ inches long with harder shells, although their flavor is equally good; recommended varieties are Colby, Greenriver, Hodge, Major, Posey and Starking Hardy Giant. Trees of at least two varieties should be planted close together, for pollination of one variety by the pollen from another is usually necessary if the trees are to bear fruit. Pecans are among the longest-lived of nut trees and may bear for 300 years or longer. A 10-year-old pecan produces about 10 pounds of nuts a year and a mature tree may bear 100 or more pounds of nuts annually.

HOW TO GROW. Pecans do best in soil with a pH of 5.5 to 6.5. For nuts within six years, buy trees 4 to 5 feet tall;

PECAN
Stuart

they should have one-year-old nut-bearing tops grafted onto branched rootstock that is several years old. Plant them in late fall or winter in Zones 6-10 and in early spring in Zone 5, cutting the trees back halfway immediately after planting. Allow several branches to develop, and when they become 4 to 5 feet long, remove all but one; this branch will serve as the trunk. Train the tree as it grows so that it will have a number of side branches spaced about 18 inches apart along the trunk; the lowest should be high enough off the ground so that you can walk beneath it. Little other pruning is needed.

Early in the spring of the second year scatter 1 pound of 10-10-10 fertilizer over the area covered by the branches. Each year increase the amount of fertilizer by 1 pound; large trees should receive 5 to 6 pounds per inch of trunk diameter annually.

Nuts should be gathered in fall as soon as possible after they drop to the ground, before the squirrels can harvest them; sometimes it may be necessary to knock or shake the nuts from the trees. Store them in a cool, rodent-proof place until used. If shelled and then stored in an airtight container in a freezer, pecans keep for as long as two years.

W
WALNUT

Black walnut (*Juglans nigra*); English walnut, also called Persian walnut (*J. regia*); butternut, also called white walnut (*J. cinerea*)

Walnuts are valuable not only as fine nut producers but also as shade trees. They generally grow fast when young and do best when set out as 3- to 4-foot trees; they become about 20 feet tall in six to eight years. Female flowers are generally inconspicuous, but male flowers are slender scaly green clusters, or catkins, about 3 inches long. Plant at least two different varieties since pollination of one variety by the pollen of another is usually necessary if the trees are to bear fruit. Since the pollen is carried by the wind, trees should be planted within 50 feet of one another. The roots of walnut trees secrete a substance known as juglone that poisons certain other plants, especially tomatoes, rhododendrons and azaleas. Do not plant vegetable or flower gardens closer than 80 feet to the trees.

The black walnut grows wild from western New England to Minnesota and Nebraska and south to the Gulf of Mexico. Cultivated varieties—such as Adams, Allen, Beck, Elmer Myers, Grundy, Snyder, Sparrow, Stambaugh and Thomas—grow in Zones 5-9, becoming 60 to 80 feet tall with an equal spread. The nuts are noted for their rich, oily flavor and, since they do not lose their flavor or texture in cooking, are highly prized in cakes, candies and ice creams. The nuts of wild trees are 1 to 1½ inches long and have a hard, thick shell; the nuts of named varieties are about 2 inches long and have softer, easier-to-crack shells. Named varieties begin to produce nuts four or five years after planting. Nuts ripen in early autumn and frequently remain on the trees for one or two weeks after the leaves have fallen. Trees may live for 100 years or more, and a mature tree produces 1 to 3 bushels of nuts annually.

The English walnut, one of the best known of nut trees, grows in Zones 5-9, but varieties differ in their adaptability to climates. Good choices east of the Rocky Mountains are Broadview, Colby, Hansen, Lake and Metcalfe, and the Carpathian variety from the mountains of Poland. For the northern part of the West Coast good varieties are Concord, Eureka, Franquette and Mayette; for southern parts of the West Coast recommended varieties are Carmelo, Drummond, Payne and Placentia. The trees become 40 to 60 feet tall with an equal spread and usually begin to bear

For climate zones and frost dates, see maps, pages 148-149.

BLACK WALNUT
Thomas

ENGLISH WALNUT
Carpathian

1½- to 2-inch easy-to-shell nuts four to seven years after planting. Trees live 60 or more years; a mature tree bears 6 bushels of nuts annually.

The butternut, or white walnut, is the hardiest of nut trees and grows well in Zones 4-8, becoming 30 to 50 feet tall with an equal spread. Although it closely resembles the black walnut, the butternut is lower and more spreading in its form. The name butternut is applied to this species because of the nuts' 1½- to 2-inch oval, oily, sweet-tasting kernels that lie within the hard, deeply ridged shell. Wild trees produce fine nuts, but a number of named varieties are especially valuable because they have larger, thinner-shelled nuts and because the trees begin to bear earlier, within two to four years after being planted. Good varieties are Craxezy, Lingle, Love, Sherwood, Thill, Van de Poppen and White. Butternut trees are short-lived, lasting 50 to 75 years; a mature tree produces 1 to 5 bushels of fruit each fall.

HOW TO GROW. Black walnuts and English walnuts do well in almost any soil, adapting to a wide range of pH, from 6.0 to 8.0, but butternuts are more sensitive to alkalinity and do best in soil with a pH of 6.0 to 7.0. Black walnuts should be planted as soon as the ground can be worked in spring, and butternuts can be planted in early spring or fall. English walnuts can be planted in early spring in Zones 5 and 6; in Zones 7-9, the trees should be planted in the fall or in winter.

Early the second spring scatter 1 pound of 10-10-10 fertilizer beneath the branches of all types of walnuts. Each year increase the amount of fertilizer by 1 pound; large trees should receive 5 to 6 pounds of 10-10-10 fertilizer per inch of diameter annually.

On the West Coast where irrigation is necessary, it is important to keep water from standing around the tree trunks. Build a shallow, permanent earthen basin under the trees with a slight mound close to the trunks to keep the bases of the trunks dry.

Cut the top of walnut trees back about halfway at the time of planting. As the tree grows, prune off branches to leave a single trunk and gradually remove the lower limbs so that there is room to walk beneath the branches. Train the tree to have three to five main branches, well spaced, 8 to 12 inches apart, and forming angles greater than 45 degrees with the trunk. Little pruning is necessary thereafter except to remove deadwood and crowded or crossing branches. Pruning should be done in the summer or fall because trees bleed heavily if they are cut in the spring.

Walnuts will fall to the ground by themselves as they ripen in the fall. When they fall, however, they should be gathered and husked promptly, then spread thinly in a shady place to dry. The kernels of walnuts have two coverings: a bony, usually ridged brown shell immediately surrounding the kernel and a pulpy, green outer hull, or husk. The husks of English walnuts fall free from the nuts when they ripen, but those of butternuts and black walnuts adhere very tightly to the shells inside them, and since these husks contain a material that will stain hands brown, it is best to wear rubber gloves when handling them. One way to loosen the husks of butternuts and black walnuts sounds like a strange and extreme method but it works: put the nuts on the driveway and drive your car back and forth over them. (This may, however, stain a light-colored driveway.) Store the husked but unshelled walnuts in plastic bags in a cool, dry, rodent-proof place; they will keep for about a year. Walnuts are easier to crack if they are soaked overnight in water to soften the shells; after the kernels are removed from their shells they will keep for at least a year if stored in airtight containers in a freezer.

BUTTERNUT
Craxezy

Herbs

A

ANISE
Pimpinella anisum

Anise is an annual cultivated for its licorice-flavored seeds, which are used after drying. The plant grows 18 to 24 inches tall and has small lacy leaves; in midsummer it bears 2-inch clusters of tiny white flowers that are followed by small gray seeds in late summer. A single plant produces one to six clusters, each bearing 6 to 10 seeds.

HOW TO GROW. Anise needs full sun and a well-drained soil with a pH of 5.5 to 6.5. Since the plants require a growing season of about four months, sow seeds as soon as possible in the spring after danger of frost is past, setting them ½ inch deep. Mix in a few radish seeds so that the quick-sprouting radish plants will mark the rows until the anise seedlings are visible. When the seedlings are about 2 inches high, remove smaller ones so that the plants stand about 8 inches apart. During the summer pull soil up around the weak stems to help support them. When the seed heads turn pale gray, harvest by cutting them off.

ANISE

B

BALM, LEMON
Melissa officinalis

Lemon balm is a perennial whose lemon-scented leaves are used as a seasoning as well as in herb tea. Lemon balm can be grown in all parts of the country, although with some difficulty in southern Florida and along the Gulf Coast, becoming 1 to 2 feet tall with a 2-foot spread and producing small pale yellow buds that open into tiny white blossoms in the summer. An average mature plant produces 2 cups of fresh leaves when it is first harvested; it may be cut at least twice during a season, although later crops will be less bountiful than the first.

HOW TO GROW. Lemon balm flourishes in full sun or partial shade and does best in soil that has a pH of 6.0 to 8.0 and is rather dry and infertile; such conditions intensify the scent of the foliage and limit growth. Sow seeds indoors in late winter and then transplant the seedlings to the garden after all danger of frost is past, or sow the seeds outdoors in late fall to germinate the following spring. Because the seeds are tiny they should be placed on top of the soil without being covered. When the seedlings are about 2 inches high, cut or pinch out the smaller ones so that plants stand 18 inches apart. The easiest way to get more plants is to divide and reset roots in early spring or to make stem cuttings during the spring or summer. Leaves can be picked from spring until late fall for use fresh; for drying, cut off stems above the second set of leaves in midsummer—the remaining leaves build strength for the next year's growth. Small plants can be potted before the first fall frost and kept indoors in a sunny window during the winter.

LEMON BALM

BASIL, SWEET
Ocimum basilicum

Sweet basil is a 15- to 24-inch bushy annual grown for its smooth clove-scented leaves. There are several forms of basil with green leaves of different sizes, and one variety, Dark Opal, has purple leaves; all, however, have a similar flavor and bear spikes of small purplish or white flowers in midsummer. A single mature basil plant will produce 3 cups of fresh leaves in a season.

HOW TO GROW. Basil needs full sun and grows best in light, sandy soil of pH 5.5 to 6.5 enriched with compost or manure. Sow seeds indoors about two months before the last frost is due, setting them no more than ½ inch deep.

SWEET BASIL

For climate zones and frost dates, see maps, pages 148-149.

After frost danger is past, move the seedlings to the garden, spacing them 12 inches apart. Or sow seeds outdoors when the weather becomes warm, and when the seedlings are 3 inches high, cut or pinch out the smaller ones so that plants stand 12 inches apart. After the plant is 4 to 6 inches tall, leaves may be picked at any time for use fresh; for drying, harvest them just as the first flowers appear by cutting off the stems about 4 to 5 inches from the ground. To produce a second, or even third, crop after cutting the stems, apply 5-10-5 fertilizer at the rate of 3 ounces per 10 feet of row. For fresh basil in winter, plant seeds in pots in late summer and keep them on a sunny window sill.

BORAGE
Borago officinalis

Borage is an annual whose fresh blossoms are floated on fruit juice or wine to impart a cooling cucumberlike fragrance. The ½-inch blue or white flowers, which bloom in summer on the 1- to 2-foot plants, can also be candied. Its 3- to 4-inch leaves are covered with bristly white hair when mature, but when young and tender are good in salads. A single young borage plant will produce about 10 flowers and 2 cups of leaves for the first harvest. Leaves may be harvested again in one month.

HOW TO GROW. Borage needs full sun and grows best in dry, rather poor soil with a pH of 6.0 to 7.0. Sow seeds outdoors in fall or very early spring, barely covering them with soil. When the seedlings are about 2 inches high, remove smaller ones so that the plants stand 12 inches apart. A few seeds scattered in a corner of the garden or on a sunny bank can be left to take care of themselves; plants will come up year after year from seeds dropped by the previous year's plants. Seeds will keep for about eight years.

BURNET
Sanguisorba minor, also called *Poterium sanguisorba*

Burnet is a bushy perennial that grows 12 to 24 inches tall and produces nearly evergreen, fernlike foliage having a cucumber flavor. It can be grown in all parts of the country, although with some difficulty in southern Florida and along the Gulf Coast. A single mature burnet plant produces ½ cup of fresh leaves for the first harvest and ¼ cup each week thereafter.

HOW TO GROW. Burnet needs full sun and does best in soil with a pH of 6.0 to 8.0. Sow the seeds ½ inch deep outdoors in late fall or early spring; when the seedlings are about 2 inches high, remove smaller ones so that plants stand 12 to 15 inches apart. Leaves should be picked when young; cut the flower stems off to encourage the development of new leaves. Burnet often springs up spontaneously from seeds dropped by mature plants, and these seedlings may be transplanted with a ball of earth.

C
CARAWAY
Carum carvii

Caraway is a biennial cultivated for its licorice-flavored seeds. Plants make a mound of feathery dark green leaves about 8 inches tall the first summer and remain practically evergreen through the winter. The following spring they send up flower stalks about 2 feet tall topped by clusters of tiny white flowers that mature by midsummer into brown seeds. The plants die after the seeds ripen. Caraway can be grown in all parts of the country, although with some difficulty in southern Florida and along the Gulf Coast. A single plant will produce about four clusters, each containing about 1 tablespoon of seeds.

HOW TO GROW. Caraway grows in full sun in almost any

BORAGE

BURNET

well-drained soil with a pH of 6.0 to 7.0. Sow the seeds about 1⁄8 inch deep outdoors in fall or early spring. When the seedlings are about 2 inches high, cut or pinch off the smaller ones so that the plants stand 6 to 12 inches apart. Harvest the seeds when they turn brown in midsummer.

CHERVIL
Anthriscus cerefolium

Chervil is an annual whose leaves resemble parsley in appearance and flavor. It becomes 1½ to 2 feet tall within eight weeks after planting and then bears clusters of minute white flowers followed by slender black seeds. One plant will produce ¼ cup of fresh leaves when first harvested, and cuttings can be made at one-month intervals if only the top leaves are removed each time.

HOW TO GROW. Chervil grows best in light shade in any soil with a pH of 6.0 to 7.0. In most parts of the country, sow seeds outdoors at three- to four-week intervals from early spring through fall; seeds sown in the fall will produce small plants that live through the winter to provide fresh leaves early in the spring. In areas where summer temperatures remain over 90° for prolonged periods, sow seeds from fall through early spring. Seeds should be barely covered with soil. When the seedlings are about 2 inches high, remove the smaller ones so that the plants stand 6 to 9 inches apart. Leaves may be picked at any time for use fresh, and pinching back will encourage new growth. Harvest chervil for drying or freezing before the flowers open, cutting the stems off at the ground. For winter use, chervil can also be started from seeds in pots indoors.

CHERVIL, GIANT See Cicely
CHINESE PARSLEY See Coriander

CHIVE
Allium schoenoprasum

Chives are bulbs of the onion family, and their hollow grasslike leaves have a mild onion flavor. They live in the garden for many years, producing plants 8 to 12 inches tall. In the spring ¾-inch clusters of purple flowers appear a few inches above the leaves. A clump 5 inches in diameter will produce 3 to 5 cups of leaves at the first cutting and can be harvested repeatedly.

HOW TO GROW. Chives do best in slightly acid soil, pH 6.0 to 7.0, that is enriched with compost or manure, but they will grow almost anywhere. Started from seeds, chives require about a year to produce; the most common way to grow them is to plant root clumps, setting them about 2 inches deep and 5 to 6 inches apart. In most parts of the country, plant in spring; but where summer temperatures remain over 90° for prolonged periods, plant the root clumps in late summer or fall to harvest during the winter or spring. Leaves can be cut at any time and picking encourages fresh growth. If plants become weak, scratch a light dusting of 5-10-5 fertilizer into the soil. Chives multiply so rapidly that the roots must be divided and reset every two or three years. For fresh chives during the winter, pot a few clumps in fall and leave them outdoors at freezing temperatures a month before bringing them in.

CICELY, SWEET (GIANT CHERVIL)
Myrrhis odorata

Sweet cicely is a thick-rooted perennial with licorice-scented leaves and seeds that are used in cooking; an oil extracted from the seeds flavors the liqueur Chartreuse. Plants grow 2 to 3 feet tall and in early summer bear flat-topped clusters of small lacy white flowers that ripen into shiny black seeds in midsummer. Sweet cicely can be grown

CARAWAY

CHERVIL   CHIVE

For climate zones and frost dates, see maps, pages 148-149.

in all parts of the country, although with some difficulty in southern Florida and along the Gulf Coast. A single mature plant will produce about 4 cups of fresh leaves and ½ cup of seeds over the growing season.

HOW TO GROW. Sweet cicely needs partial shade and does best in a moist soil of pH 5.5 to 6.5 supplemented with compost or manure. Sow seeds in late summer or early fall; they then sprout in spring. Plant seeds ¼ inch deep; in spring when seedlings are about 2 inches high, remove the smaller ones so that the plants stand 2 feet apart. Because the leaves remain green from very early spring until late fall, sweet cicely leaves are almost always used fresh. Harvest seeds when they turn black. Established plants can be divided for propagation in fall.

CORIANDER (CHINESE PARSLEY)
Coriandrum sativum

Coriander is an annual whose lemony-flavored seeds are widely used in baking and in spice mixtures, such as curry powder. The fresh leaves, though too pungent for many tastes, are popular in the Orient (hence the name Chinese parsley) and in Latin America (where the herb is known as *cilantro*). Coriander plants grow 12 to 30 inches tall and have finely divided fernlike foliage. In midsummer the stems are topped by clusters of tiny white, pink or lavender flowers that ripen in late summer into seeds resembling peppercorns. A single coriander plant will produce ¼ cup of seeds over a two-month period or, if the flowers have been removed to prevent the formation of seeds, ½ cup of leaves in a similar period.

HOW TO GROW. Coriander needs full sun, but will grow in almost any soil with a pH of 6.0 to 7.0. Choose a location sheltered from the wind. Sow the seeds outdoors in the spring after danger of frost has passed, setting them ¾ inch deep. When the seedlings are about 2 inches high, remove the smaller ones so that plants stand 8 to 10 inches apart. Leaves may be picked at any time. Because the seeds easily fall off the plant, it is necessary to harvest them carefully: when the seeds begin to turn brown, cut the entire plant off and drop it into a bag so that the seeds can be removed later. Seeds keep for about five years.

D
DILL
Anethum graveolens, also called Peucedanum graveolens

Dill is an annual cultivated for its seeds and leaves, which are often used interchangeably, as well as for its flower heads, which are a pickle seasoning. Plants grow about 3 feet tall and spread to an equal distance. In midsummer they bear flat-topped clusters of tiny yellow flowers that ripen into small flat seeds. A single plant produces ¼ cup of seeds, ½ cup of leaves and four to eight flower heads. Only rarely can it be harvested more than once.

HOW TO GROW. Dill needs full sun and does best in a soil of pH 5.5 to 6.5 supplemented with compost or manure. Choose a spot protected from the wind because otherwise the plants may blow over or seeds may blow away as they begin to ripen. Sow seeds outdoors in early spring or just before the ground freezes in the fall, setting them ¼ inch deep. When the seedlings are about 2 inches high, remove the smaller ones so that plants stand 12 inches apart. If growth seems weak when the plants are about 1 foot tall, scatter 5-10-5 fertilizer around them at the rate of 3 ounces to 10 feet of row. Pick the leaves at any time for use fresh. For drying, pick leaves early in summer and chop them. Harvest seeds when they begin to turn brown: cut the entire plant off at the ground and drop the plant into a paper bag so that the seeds can be removed later.

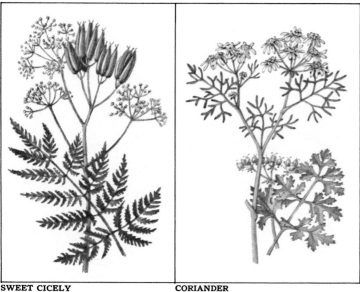

SWEET CICELY CORIANDER

DILL

F

FENNEL
Foeniculum vulgare, also called *F. officinale*

Fennel, a perennial usually grown as an annual, is valued for its licorice-flavored leaves and seeds. The plant grows 3 to 4 feet high; its green, feathery foliage is topped in summer by small clusters of tiny yellow flowers that are followed by brown seeds in late summer. A single fennel plant will produce 1 cup of leaves and ¼ cup of seeds during its growing season. It may be cut at least twice in the summer. Common fennel should not be confused with the related sweet, or Florence, fennel *(F. vulgare dulce)* or carosella fennel *(F. vulgare piperitum),* both of which are thicker-stemmed plants, grown and used as vegetables.

HOW TO GROW. Fennel needs full sun and is easy to grow in any soil with a pH of 6.0 to 8.0. Sow the seeds outdoors in early spring, barely covering them with soil; successive sowings can be made at 10-day intervals until early summer. When the seedlings are about 2 inches high, remove the smaller ones so that plants stand about 12 inches apart. When plants are about 1 foot tall, feed them with 5-10-5 fertilizer at the rate of 3 ounces to 10 feet of row. Leaves may be picked at any time; harvest the seeds when they begin to turn brown.

G

GARLIC
Allium sativum

Garlic is a relative of the onion that lives in the garden for many years and is grown for its pungent-tasting bulb. Each plant produces a single bulb made up of 8 to 12 claw-shaped sections called cloves, which are wrapped together by a parchmentlike covering. Garlic plants have slender straplike leaves that grow 1 to 2 feet tall; during early summer they send up slender flower stalks topped by small globular clusters of tiny white flowers. Sometimes tiny, but edible, aboveground bulbs appear among the flowers.

HOW TO GROW. Garlic needs full sun and does best in a light, sandy soil of pH 5.5 to 8.0 supplemented with compost or manure. Grow garlic from cloves planted in early spring, setting them 1 inch deep and 6 inches apart. When the plants reach 6 inches, apply 5-10-5 fertilizer at the rate of 3 ounces to 10 feet of row and scratch it into the soil. Harvest garlic by digging up the plants when the foliage dies down in the early fall. To speed ripening, bend the tops of the plants over to the ground late in summer. After the plants are harvested, the leaves can be braided so that the bulbs can be hung together.

GIANT CHERVIL See Cicely

L

LOVAGE (SMALLAGE)
Levisticum officinale

Lovage is a perennial whose celery-flavored leaves and seeds are used for seasoning, while its stems are eaten like celery. The plant forms a 2- to 3-foot mound of shiny, dark green leaves. In early summer it sends up hollow, branched stalks as tall as 7 feet; they carry clusters of small greenish yellow flowers that ripen into fawn-colored seeds in late summer. Lovage can be grown in any part of the country, although with some difficulty in southern Florida and along the Gulf Coast. A single mature lovage plant in its second year will produce ½ cup of seeds and 4 cups of leaves, and six to eight stalks in a season.

HOW TO GROW. Lovage will grow in full sun or partial shade; it does best in a moist soil of pH 6.0 to 7.0 that is supplemented with compost or manure. Seeds may be sown

FENNEL

GARLIC

For climate zones and frost dates, see maps, pages 148-149.

143

LOVAGE

SWEET MARJORAM   MINT

indoors in spring in 3- to 4-inch peat pots and set 3 feet apart in the garden when night temperatures remain about 40°. Alternatively, sow seeds outdoors in early fall, setting them ¼ inch deep. Plants may also be started by dividing and resetting roots in spring. Feed lovage in the spring by scattering a handful of 5-10-5 fertilizer around each plant and scratching it into the soil. Leaves can be picked at any time for use fresh; to pick them for drying, cut the stems above the second set of leaves—the remaining leaves build strength for the next year's growth.

M

MARJORAM, SWEET
Origanum majorana, also called *Majorana hortensis*

Sweet marjoram is a perennial, but it is easier to grow as an annual. Its aromatic leaves and tender stem tips are used to season foods, and its oil is used as a perfume in soapmaking. Plants grow about 12 inches tall with an equal spread and bear tiny white flowers in midsummer. A single plant will produce ½ cup of leaves when it is first harvested and ¼ cup of leaves at the second cutting.

HOW TO GROW. Sweet marjoram needs full sun and does best in a light, sandy soil with a pH of 6.0 to 8.0. Plants can be started from tiny seeds sown outdoors in the spring; alternatively, plant them indoors in late winter and set into the garden after the danger of frost has passed, spacing them 8 to 15 inches apart. Leaves and stem tips may be picked at any time for use fresh. When the plants blossom, harvest the leaves for drying by cutting the stems off above the second set of leaves. New stems will come up after the first ones are cut so that two, and sometimes three, crops of leaves can be harvested in a single season. For fresh sweet marjoram during the winter months, put a few young plants in pots in the fall and place them on a sunny window sill indoors.

MINT
Apple mint, round-leaved mint or woolly mint *(Mentha rotundifolia);* orange mint, also called bergamot mint *(M. citrata,* also called *M. odorata);* black peppermint *(M. piperita vulgaris);* white peppermint *(M. piperita officinalis);* spearmint *(M. spicata)*

The mints listed here are perennials cultivated for their aromatic leaves. They can be grown in any part of the country, although with some difficulty in southern Florida and along the Gulf Coast.

Apple mint grows ½ to 3 feet tall and has nearly round, soft-textured gray-green leaves that are hairy above and densely covered with a white "wool" beneath. It bears 2- to 4-inch spikes of purplish white flowers in the summer. Orange mint, which has a citrus-mint flavor, grows about 12 inches tall; it has reddish green stems and smooth, rounded or heart-shaped dull green leaves with purple blotches. It bears short spikes of reddish purple flowers in the summer. Black peppermint grows 2 feet tall and has purplish stems set with 3-inch narrow dark green leaves. It bears small spikes of purple flowers in the autumn. The leaves yield an especially fragrant oil used in mentholated products, chewing gum and confections. White peppermint, whose oil is used in making crème de menthe, grows 1 to 2 feet tall and has reddish stems and 1- to 2-inch narrow green leaves. It bears small spikes of purple flowers in late summer. Spearmint is the most intensely flavored of the mints. It grows 1 to 2 feet tall with sawtooth-edged glossy leaves about 2 inches long and bears violet or pink flower spikes in midsummer. A mature mint plant yields 2 cups of fresh leaves at the first cutting and smaller amounts at 10-day intervals thereafter.

HOW TO GROW. Mints grow best in full sun or light shade and do best in moist soil of pH 5.5 to 6.5 supplemented with compost. The easiest way to start a mint bed is to buy plants and set them into the garden in spring, spacing them about 12 inches apart. Each spring chop the soil with a sharp spade to a depth of 3 to 4 inches to break up the matted roots, and feed the plants with a dusting of 5-10-5 fertilizer. Do not use manure; it sometimes carries mint rust fungus, a disease that disfigures and may kill the foliage. Because the roots spread rapidly, the bed should be enclosed with an underground barrier, such as the metal or plastic strips used to edge a flower bed; and the roots should be divided and reset every three years in spring.

Mint leaves can be picked at any time for use fresh, and picking will force new growth. Harvest leaves for drying just as the flowers begin to open, cutting the stems above the first two sets of leaves. For fresh mint in winter, set a few plants in pots in early fall and place them on a sunny window sill indoors.

P

PARSLEY
Petroselinum crispum, also called *P. hortense*

Parsley is a biennial cultivated for the leaves that it produces during its first year when the plant sends up a foliage tuft 8 to 12 inches tall and of equal spread. If not dug up, the plant will send up a 2-foot stalk during its second year and bear clusters of tiny greenish flowers that produce brown seeds. The most popular kind of parsley is that with curly foliage; good varieties are Champion Moss Curled and Perfection. If flat-leaved parsley is desired, Plain Italian Dark Green is a good variety. A single mature plant of either type produces about 1 cup of leaves in three weeks. A third type, Hamburg, or Turnip-Rooted, parsley, is grown principally for its 8- to 10-inch white roots, which are used as a vegetable, but its leaves are also good for the same purposes as those of other types.

HOW TO GROW. Parsley grows best in full sun, but will tolerate light shade. The soil should be enriched with compost or manure and have a pH of 5.0 to 7.0. In most parts of the country, sow seeds in very early spring or in late fall just before the ground freezes; for spring sowing, soak seeds in tepid water for 24 hours before planting to speed germination. In areas where summer temperatures are over 90° for prolonged periods, sow seeds in fall. Plant seeds ¼ inch deep, and when the seedlings are about 2 inches high, remove smaller ones to space plants 3 inches apart. When the plants have grown enough to touch one another, pull and use alternate plants; repeat when they touch again, making the final spacing 12 inches. When the plants are 4 inches tall and again about one month later, feed with 5-10-5 fertilizer at the rate of 3 ounces per 10 feet of row. Leaves may be picked at any time. For fresh parsley in winter, set a few young plants in pots and place them on a sunny window sill indoors or in a cold frame outdoors.

PARSLEY, CHINESE See Coriander

R

ROSEMARY
Rosmarinus officinalis

Rosemary is an evergreen shrub grown for its resinous-tasting needlelike leaves, which are about ¾ to 2 inches long, dark green above and furry white beneath. Plants bear ½-inch violet-blue flowers in the spring and occasionally in other seasons. In most of the country rosemary must be grown as an indoor pot plant during the winter and as a temporary garden plant or pot plant during the

PARSLEY

ROSEMARY

For climate zones and frost dates, see maps, pages 148-149.

145

SAGE

summer. In climates where winter temperatures do not fall below 10° for prolonged periods, rosemary can be grown in the garden and may become as much as 6 feet in height, but seldom attains more than 2 to 3 feet if even occasional winter frosts nip back the top branches. Rosemary plants vary widely in size. If you cut only the tips of the stems, you should be able to harvest at least a cup at a time.

HOW TO GROW. Rosemary needs full sun and will grow in any soil with a pH of 6.0 to 7.5; a relatively poor dry soil is preferable to a fertile moist one because plants will grow more compactly and have more fragrance. Rosemary can be grown from seeds sown ¼ inch deep in packaged potting soil indoors at any time, but the plants require two or three years to mature. The most common way to grow rosemary is to buy one or two plants and propagate them from stem cuttings. Leaves can be picked at any time for fresh use; for drying, harvest leaves when flowers appear by cutting off one half of the current season's growth.

S

SAGE
Salvia officinalis

Sage is a perennial grown for its aromatic leaves. The leaves, 2 to 4 inches long, are usually gray green, but in some varieties are purple, gold-edged or variegated with pink and white. Plants become 1½ to 2½ feet tall and in spring have small lilac-blue flowers that attract bees. Sage can be grown in all parts of the country, although with some difficulty in southern Florida and along the Gulf Coast. A single mature sage plant will produce 1 cup of fresh leaves at each of two harvestings in a season.

HOW TO GROW. Sage needs full sun and grows in any soil with a pH of 5.5 to 6.5. Sage can be grown from seeds sown ¼ inch deep outdoors in spring, but the plants require about two years to reach usable size. It is easier to buy one or two plants, plant them 1½ to 2 feet apart and propagate them from root divisions in early spring or from stem cuttings in summer. For strong new growth each year, cut the previous year's growth back by half each spring. After four or five years plants become weak and should be removed to make room for seedlings. Leaves may be picked at any time for use fresh. To harvest leaves for drying, cut 6- to 8-inch pieces from the tips of the stems as the flower buds appear; a second and often a third crop can be harvested. Do not cut any stems after early fall because late new growth may suffer injury from winter cold.

SUMMER SAVORY

SAVORY, SUMMER
Satureia hortensis

Summer savory is an annual cultivated for its ½- to 1-inch tangy leaves. Plants grow 12 to 18 inches high and about two months after sowing, bear tiny pinkish flowers. A single summer savory plant will produce ¼ cup of leaves and can usually be harvested only once.

HOW TO GROW. Summer savory needs full sun but grows in any soil with a pH of 6.0 to 7.0. Sow seeds in early spring, setting them ¼ inch deep; make successive sowings at three- to four-week intervals if desired. When the seedlings are about 2 inches high, remove the smaller ones so that plants stand 4 to 6 inches apart. Leaves may be picked at any time for use fresh; to pick for drying, cut the plants off at ground level before the flowers appear.

SHALLOT
Allium ascalonicum

Shallot is a bulb of the onion family that lives for many years and is grown for its mild garlic-flavored roots, which are made up of many segments called cloves, each with its

SHALLOT

own parchmentlike cover. The bulbs have slender tubelike leaves, so that a young shallot plant resembles a bunch of scallions. The plants grow about 18 inches tall and sometimes bear tiny white or violet flowers in early summer.

HOW TO GROW. Shallots grow best in soil with a pH of 5.5. to 7.0. In most parts of the country plant the cloves outdoors in early spring to harvest in summer, setting them about 2 inches deep and 2 to 3 inches apart. Where summer temperatures remain over 90° for prolonged periods, plant the cloves in the late summer or fall to harvest during the winter and spring. When the plants are 5 to 6 inches tall, feed them with 5-10-5 fertilizer at the rate of 3 ounces per 10 feet of row. To harvest the bulbs for winter use, dig up the plants as soon as the leaves wither in fall.

SMALLAGE See Lovage

T

TARRAGON, also called FRENCH TARRAGON
Artemisia dracunculus

Tarragon is a 2- to 2½-foot-tall perennial cultivated for its 2-inch licorice-flavored leaves. It grows in all parts of the country, although with some difficulty in southern Florida and along the Gulf Coast, and is unique in that it does not produce viable seeds. A single mature tarragon plant will produce about 2 cups of fresh leaves at the first cutting and about ½ cup one month later.

HOW TO GROW. Tarragon will grow in full sun or partial shade and does best in a light, sandy soil with a pH of 6.0 to 7.5. Because many seed packets labeled "tarragon" do not contain *A. dracunculus* but another, less flavorful species called Russian tarragon, growing plants from seeds is not recommended; buy small plants instead. Plant them 2 feet apart in early spring. Feed tarragon in early spring and in early summer after the first crop of leaves is cut by scattering a handful of 5-10-5 fertilizer around each plant and scratching it into the ground. Leaves may be picked at any time for use fresh; for drying, harvest in early summer and again in early fall, cutting the stems 3 inches from the ground. Propagate from root divisions in early spring. Tarragon plants should be divided and reset every third spring.

THYME and LEMON THYME
Thyme, also called garden thyme, common thyme (*Thymus vulgaris*); lemon thyme (*T. citriodorus*)

Thyme and lemon thyme are 6- to 12-inch nearly evergreen bushes with intensely aromatic ¼-inch leaves. They can be grown in the garden in every part of the country and are often planted as an edging beside a path to scent the air. In early summer all thymes have tiny lilac-colored fragrant flowers that attract bees. A single mature thyme plant will produce ½ cup of fresh leaves at each of two harvestings in a season.

HOW TO GROW. Thyme needs full sun and a light, sandy soil with a pH of 5.5 to 7.0. Garden thyme can be grown from seeds sown ¼ inch deep outdoors in spring, but the plants require about two years to reach usable size. The easiest way to start a thyme bed is to buy a plant or two and propagate them from stem cuttings. Set plants 12 inches apart at the same depth they have been growing. Scatter a tablespoon of cottonseed or bone meal beneath each plant in early spring and scratch it into the soil; do not overfertilize because fast growth makes the plants susceptible to injury from winter cold. Plants become overgrown in three or four years and should be divided. Leaves may be picked at any time for use fresh; for drying, harvest leaves just before the flowers open by cutting off one half of the current season's growth.

TARRAGON

THYME

For climate zones and frost dates, see maps, pages 148-149.

How temperatures affect planting

The maps shown here, used with the information given in the encyclopedia, will help indicate which fruit and nut varieties will grow successfully in your region, as well as when to sow seeds or set out many vegetables and herbs.

Winter temperature is the critical factor for berries, grapes, fruit trees and nut trees, which vary in their ability to survive cold; encyclopedia entries for these plants include recommended areas, keyed by number to the zones shown on the map below. By checking the map a gardener in Milwaukee, Wisconsin, for example, can determine that he lives in Zone 5. If he wants to grow apples, he should choose varieties recommended for that zone—say, McIntosh or Cortland.

Planting dates for some vegetables and many herbs hinge on the last spring frost or the first fall frost, indicated in general terms for various parts of the country on the maps opposite. For example, early varieties of broccoli should be started indoors two to three months before the expected date of the last spring frost in a given region. Specific frost dates vary widely within each region—even within a neighborhood—so it is advisable to check with your weather bureau for more precise local figures and to keep a record of temperatures in your own garden from year to year. The selection and culture of vegetable varieties depend on certain maximum summer or minimum winter temperatures; in these cases, too, local weather bureau records are your best source of specific data to use with the encyclopedia.

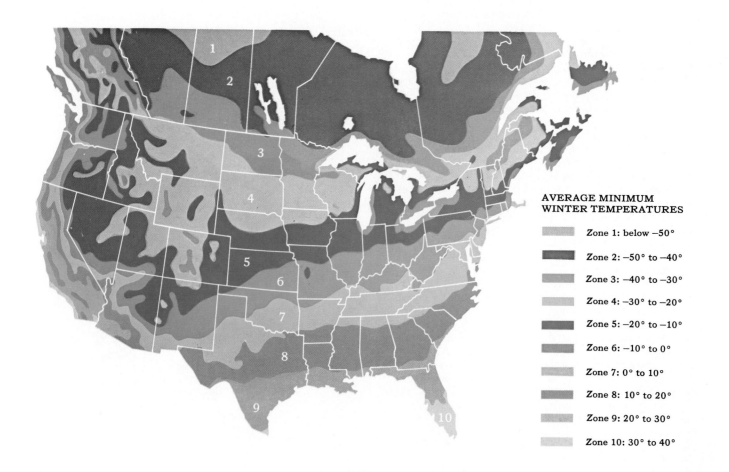

AVERAGE MINIMUM
WINTER TEMPERATURES

Zone 1: below −50°

Zone 2: −50° to −40°

Zone 3: −40° to −30°

Zone 4: −30° to −20°

Zone 5: −20° to −10°

Zone 6: −10° to 0°

Zone 7: 0° to 10°

Zone 8: 10° to 20°

Zone 9: 20° to 30°

Zone 10: 30° to 40°

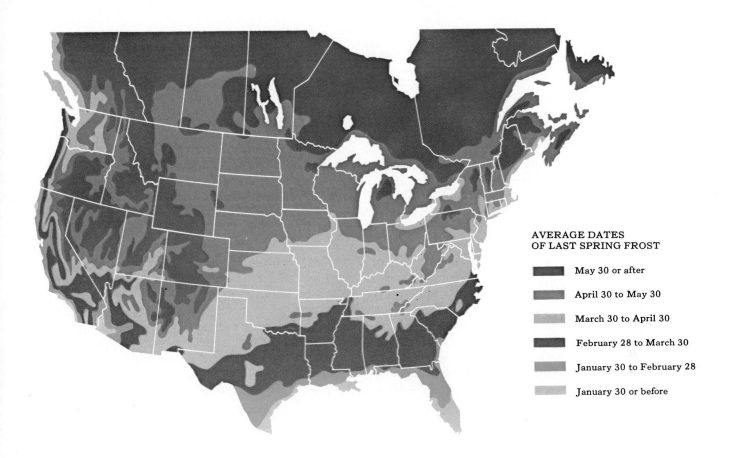

AVERAGE DATES
OF LAST SPRING FROST

May 30 or after

April 30 to May 30

March 30 to April 30

February 28 to March 30

January 30 to February 28

January 30 or before

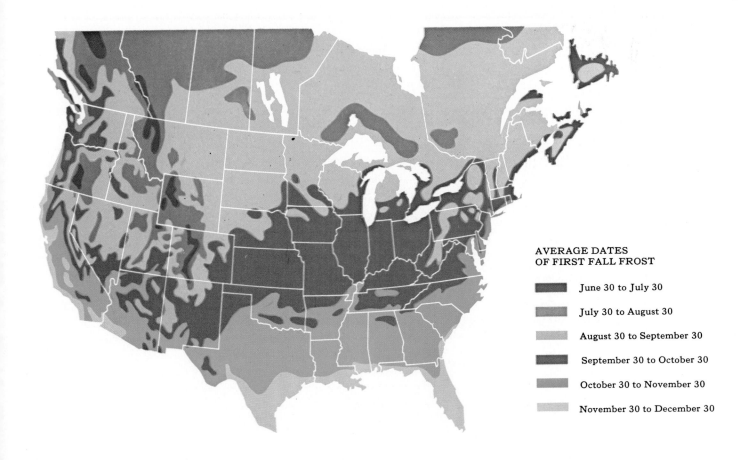

AVERAGE DATES
OF FIRST FALL FROST

June 30 to July 30

July 30 to August 30

August 30 to September 30

September 30 to October 30

October 30 to November 30

November 30 to December 30

Pest control for fruits and nuts

Although fruit trees are especially vulnerable to insects and diseases, any kind of fruit or nut plant can suffer from their attacks and by the time the damage is evident it is usually too late to save the crop. The best defense is to get an early start on control measures. Buy disease-resistant varieties of plants if possible. Keep the ground around plants cleared of weeds and debris, which harbor insects and diseases, and prune away dead, diseased or broken branches and canes, which invite the spread of trouble.

To guarantee a good crop, you will probably have to use some pesticides: how much will depend on how prevalent pests are in your area (ask your local agricultural extension service) as well as on how perfect you want your fruit. In the charts below and on the opposite page, the most important applications are printed in blue type; the others are also advisable if infestations in your area tend to be severe. Some plants, notably apple trees, need many treatments because they are the targets of many kinds of pests over long periods of time; others, like citrus fruits, which are beset by scale insects, need repeated treatment because the pests are tenacious.

Most insects and diseases can be controlled by one of the multipurpose fruit and nut sprays sold at garden centers, designated by the abbreviation MP on the charts; a typical one combines fungicides such as captan or zineb with insecticides such as malathion and methoxychlor. A dormant oil spray, made up of a highly refined grade of petroleum mixed with water, is also required for fruit trees. Citrus trees need a lighter dormant oil spray specifically designated for their use. Pay strict attention to the directions on the labels of the sprays, and be sure to discontinue spraying at the time specified by the manufacturer.

SMALL FRUITS

BLUEBERRIES **CURRANTS** **GOOSEBERRIES**	MP *Apply when buds become green, usually 2 to 3 weeks before flowering	MP Apply immediately after bloom, and twice more at 7- to 10-day intervals

BLACKBERRIES **RASPBERRIES**	MP Apply when leaves first unfold	MP Apply when new cane growth is 6 to 8 inches high	MP Apply just before flower buds open
GRAPES	MP Apply when new growth is 6 to 10 inches long	MP Apply just before flower buds open	MP Apply immediately after bloom and twice more at 14-day intervals
STRAWBERRIES	MP Apply early in spring, as soon as winter mulch is removed	MP Apply at 7-day intervals from the time new growth starts until flower buds open	MP Apply twice at monthly intervals after harvest

NUT TREES

CHINESE CHESTNUT **PECAN** **WALNUT**	MP Apply in spring when new growth is 2 to 3 inches long	MP Apply when new leaves are about half their mature size	MP Apply in midsummer when nut tips begin to turn brown	MP Apply 3 weeks later and again 4 weeks later
ALMOND	Follow the spray schedule recommended for peaches (opposite).			

CITRUS FRUITS

GRAPEFRUIT **KUMQUAT** **LEMON, LIME** **ORANGE** **TANGERINE**	MP Apply in early spring after blossom petals have fallen but before fruits have reached ¾ inch in diameter	MP Apply in midspring 4 to 6 weeks later	Dormant oil Apply in early summer 1 month later	MP Apply in midfall 3 to 4 months later

*Multipurpose spray sold for use on home-garden fruits

| DORMANT | GREEN TIP | ½ INCH GREEN | PINK | FULL BLOOM | PETAL FALL |

A spray schedule for fruit trees

To produce top-grade tree fruits, professional growers follow a complete spraying routine like the 13-step schedule shown below. Home gardeners may be willing to settle for less than perfection, but even they will probably need to apply at least the basic sprays indicated in blue in order to get an edible crop.

Because the times at which flowers and fruits form vary from year to year, spraying is scheduled not by the calendar but by key stages of blossom development, pictured above. These stages, and the type of spray to use at each one, are indicated for various types of fruit trees in the chart below; at the bottom of the chart are the insects and diseases controlled by the sprays, with the span of time each is apt to be in evidence.

At the first development stage, while trees are dormant and before their buds begin to swell, a dormant oil spray is applied in early spring, when the temperature is between 45° and 85°. It smothers many insect eggs and hibernating pests before they can get started. On certain trees—peaches, nectarines, apricots and almonds—that are susceptible to a disease known as peach leaf curl, the dormant oil treatment should be preceded by one with the fungicide called Bordeaux mixture. It can be applied at any time during the winter, but no later than a week before you apply the dormant oil. Subsequent spraying of fruit trees varies from fruit to fruit, as indicated below, until the petals fall. Thereafter all trees require a series of treatments with one of the multipurpose sprays, applied at the intervals shown.

STAGES AT WHICH TO SPRAY

FRUIT TREES	DORMANT	GREEN TIP	½ INCH GREEN	PINK	FULL BLOOM	PETAL FALL	NUMBER OF DAYS AFTER PETAL FALL						
							10	20	30	40	50	60	70
APPLES	dormant oil	captan	captan	MP*	captan	MP	MP	MP	MP	MP	MP	MP	MP
APRICOTS NECTARINES PEACHES	Bordeaux mixture; dormant oil			MP	captan	MP	MP	MP	MP	MP			
CHERRIES	dormant oil			MP	captan	MP	MP	MP					
PEARS	dormant oil			MP	Bordeaux mixture	MP	MP	MP	MP			MP	MP
PLUMS	dormant oil			MP	captan	MP	MP	MP	MP	MP			

scale insects
mite and aphid eggs
peach leaf curl
peach twig borer

PESTS AND DISEASES
AFFECTED BY SPRAYING

fire blight

peach-tree borer
apple maggot
Japanese beetle

mites

codling moth, pear psylla, plum curculio

Oriental fruit moth

aphid

apple scab (present all year)

*Multipurpose spray sold for use on home-garden fruits

Controlling vegetable pests

Many insects can be kept under control in a vegetable garden by simple nonchemical methods. Some, like leaf hoppers, can be washed off plants with a stream of water; slow-moving types, like Colorado potato beetles, can be picked off by hand. Keeping the garden free of weeds and debris, in which insects hibernate and lay eggs, will discourage Mexican bean beetles and tarnished plant bugs; fitting wax-paper plant caps or paper collars over

PEST	DESCRIPTION	METHODS OF CONTROL
	IMPORTED CABBAGE WORMS Since it arrived in Quebec in 1860 from Europe, this insect has spread throughout North America. The adult butterfly deposits eggs by the hundreds on the undersides of leaves in spring. Velvety green worms hatch a week later and begin eating the leaves; when an inch long, they spin cocoons, emerging later as butterflies to begin the cycle again. The worms may reproduce five or six or more times a season. SUSCEPTIBLE PLANTS: BRUSSELS SPROUTS, CABBAGE, CAULIFLOWER, COLLARDS, KOHLRABI, RADISH, TURNIP	If worms appear, dust plants with *Bacillus thuringiensis,* an insect-killing microbe. Or apply rotenone or carbaryl, covering both sides of leaves. For serious infestations, use malathion.
	STRIPED CUCUMBER BEETLES These beetles—less than ¼ inch long—feed on young plants; wormlike larvae from their eggs eat the roots and in summer become adults that chew holes in leaves, flowers, and cucumber and melon rinds. They also spread bacterial wilt. SUSCEPTIBLE PLANTS: BEAN, CANTALOUPE, CORN, CUCUMBER, PEA, SQUASH, WATERMELON	Start seeds or seedlings under wax-paper plant caps; push the caps 1 inch into the ground to keep out the egg-laying beetles. If beetles appear, apply rotenone or methoxychlor or carbaryl.
	MEXICAN BEAN BEETLES Mexican bean beetles are voracious pests less than ¼ inch long that emerge from hibernation in spring. Adult beetles, and later their larvae, attack leaves from underneath, reducing them to lacy skeletons. If unchecked the beetles will go on to devour bean pods and stems as well. SUSCEPTIBLE PLANTS: ALL TYPES OF BEANS	Treat infested foliage with rotenone or malathion, carbaryl or diazinon, covering the undersides of leaves. After harvest, clean up plant debris in which beetles could hibernate.
	COLORADO POTATO BEETLE This pest, native to the Rocky Mountains, has spread across North America to Europe. The ⅜-inch-long beetles winter underground, emerging when plants are up to lay their eggs on the undersides of leaves. If unchecked, the beetles and their fat red grubs destroy foliage and kill the plants. SUSCEPTIBLE PLANTS: EGGPLANT, PEPPER, POTATO, TOMATO	Knock the beetles or larvae off by hand into a jar or pail of water coated with a film of oil or kerosene, or treat the plants with carbaryl or diazinon.
	SPIDER MITES These almost microscopic pests may be red but are usually yellow, brown or green and are nearly invisible where they congregate on the undersides of leaves to suck sap. The first sign of infestation may be white flecks on leaves, followed by the white mealy cobwebs that the pests spin from leaf to leaf. Hatching and maturation accelerate in hot, dry weather, so that mites are generally most abundant in summer. SUSCEPTIBLE PLANTS: BEAN, CUCUMBER, EGGPLANT, MELON, TOMATO	Knock mites off the leaves with a stream of water, using enough force to break the webs. For serious infestations treat most vegetables at 10- to 14-day intervals with diazinon or dicofol. On eggplants, use malathion instead of dicofol.
	ROOT MAGGOTS Root maggots, like the ⅓-inch-long cabbage maggot shown, are the larvae of flies that hibernate in the soil over winter and emerge in spring or early summer to lay eggs on the stems or leaves of young plants. After hatching, the larvae go underground to feed on roots. Working invisibly, they can stunt the plants or destroy an entire crop of root or bulb vegetables by riddling the roots with tunnels. SUSCEPTIBLE PLANTS: BROCCOLI, CABBAGE, CARROT, CELERY, CORN, ONION, PARSNIP, PEA, RADISH, TURNIP	When sowing seeds or planting seedlings, soak the ground around the plants with diazinon. Or, to prevent the hatching larvae from reaching the roots, slit a 3-inch square of tar paper and slip it around the stem of each plant at ground level. After harvest, destroy any damaged plants.

young seedlings will protect them from ground attack by cut-worms and striped cucumber beetles. But you probably will have to spray or dust with chemicals to control serious attacks as described below. Carefully follow the instructions on the label; it may be necessary to stop applications days or even weeks before harvesting, depending on the type of vegetable and the pesticide, and it is always advisable to wash vegetables well.

PEST	DESCRIPTION	METHODS OF CONTROL
LEAF HOPPERS	When the plants on which they feed are disturbed, these ⅛-inch-long wedge-shaped insects jump or hop from leaf to leaf. They collect on the undersides of leaves, sucking the sap until the foliage yellows or its edges become brown. Leaf hoppers spread aster yellows, a virus disease that stunts growth, causes vegetables to ripen prematurely or develop a warty skin, and may cause plants to yellow and die. SUSCEPTIBLE PLANTS: BEAN, CELERY, EGGPLANT, LETTUCE, POTATO, TOMATO	Knock leaf hoppers off the plants with water from a garden hose. Apply methoxychlor or rotenone. For severe infestations, use malathion or carbaryl.
EUROPEAN CORN BORERS	These 1-inch-long caterpillars feed chiefly inside plant stems and the first sign of trouble may be broken or hole-riddled stalks or—in the case of corn—broken tassels. In corn they eventually eat into the ears. Borers spend the winter in old stems and emerge as yellowish moths in early spring, flying about at night to deposit masses of eggs on the undersides of young leaves. By late spring, young borers begin to hatch and eat holes in the leaves before boring into stalks. SUSCEPTIBLE PLANTS: BEAN, BEET, CELERY, CORN, PEPPER, POTATO	As soon as corn tassels are visible, or when other vegetables are about half grown, treat the plants with carbaryl, repeating the application three more times at intervals of five days.
APHIDS	Among the most common of all garden pests, aphids are less than ⅛ inch long and may be black, green, red, yellow, lavender, brown or gray. They attack leaves and stems, and by sucking sap they make leaves curl or pucker. Aphids spread virus diseases and excrete a substance called honeydew, which attracts a fungus known as sooty mold. SUSCEPTIBLE PLANTS: ALL VEGETABLES	Knock aphids off the plants with a stream of water from a garden hose. For severe infestations, apply nicotine sulphate, rotenone or malathion weekly.
CUTWORMS	The most common type of cutworm eats its way through plant stems, toppling the plants. The worms winter underground to emerge in spring, feeding at night and spending the day coiled beneath the soil. Most of them, like the 2-inch black cutworm shown, spin cocoons in summer and the moths appear a few weeks later to lay new eggs, but some Southern species reproduce several times a year. SUSCEPTIBLE PLANTS: ALL VEGETABLES, ESPECIALLY NEWLY TRANSPLANTED SEEDLINGS OF ASPARAGUS, CABBAGE, PEPPER, TOMATO	Set seedlings inside paper collars (*drawings, page 26*) to prevent the cutworms from reaching the plants, or control them by spraying plants with diazinon or carbaryl.
TARNISHED PLANT BUGS	These ¼-inch-long pests, named for the tarnishlike blotches on their backs, inject plants with a substance that blackens and deforms the tips of leaves and the joints of stems. The bugs hibernate under dead leaves, weeds or stones and emerge in spring; their life cycle is only three to four weeks long, but they produce up to five generations in a season. SUSCEPTIBLE PLANTS: BEAN, BEET, CABBAGE, CAULIFLOWER, CELERY, CHARD, CUCUMBER, POTATO, TURNIP	At the first sign of blackening leaf tips, apply rotenone and repeat at weekly intervals. After harvest, clean up weeds and garden debris in which the bugs could hibernate.

Diseases afflicting vegetables

Like vegetable pests, vegetable diseases can be controlled by chemicals, but the best measures are preventive. Select disease-resistant varieties; use fungicide-treated seeds; rotate crops by planting in different areas of the garden; and destroy infected plants. If you do use a fungicide, follow the directions on the label and stop its use before harvesting at the time specified.

DISEASE	DESCRIPTION	METHODS OF CONTROL
POWDERY MILDEW	This fungus, which thrives in cool, humid weather, is first seen as whitish spots on the lower surfaces of leaves. Eventually leaves and stems become coated with patches of the powdery growth, and the leaves may turn brown and shrivel. Vegetables may ripen prematurely, be malformed and suffer sunscald caused by the lack of any foliage cover. SUSCEPTIBLE PLANTS: BEAN, CANTALOUPE, CUCUMBER, PUMPKIN, SQUASH, WATERMELON	Choose resistant varieties. At the first sign of powdery mildew, apply benomyl, dinocap or sulfur; repeat the application in seven to 10 days.
EARLY BLIGHT	Infection appears in early summer, first on lower, shaded leaves as dark brown spots; these often develop into concentric rings in bull's-eye patterns. Severely infected leaves turn yellow, then brown and fall early, starting at the base of the plant. Dark sunken spots often appear at the stem ends of ripening fruits. Heavy infection will seriously diminish production of the fruits. SUSCEPTIBLE PLANTS: POTATO, TOMATO	Apply maneb, zineb, chlorothalonil or mancozeb when tomatoes first appear, and repeat weekly throughout the growing season. Apply to potatoes every seven to 10 days after the plants grow to be 6 inches high.
RUST	Slightly raised whitish pustules appear on the undersides of leaves. Within a few days the pustules produce powdery red or brown spores that are spread by water or wind, infecting other plants. The leaves turn yellow, then brown and die. Plants are stunted or mature early and yield is reduced. Rust is most common in areas of high humidity. SUSCEPTIBLE PLANTS: ASPARAGUS, BEAN	Select rust-resistant varieties. Since rust spores are spread by moisture, water early in the day for fast evaporation and do not walk among the plants when they are wet. If rust appears, apply maneb or zineb to asparagus every 10 days, to beans at weekly intervals.
DAMPING-OFF	Damping-off fungi attack young seedlings at the soil line, causing them to wilt, then topple over and die. SUSCEPTIBLE PLANTS: ALL SEEDLINGS	Start seeds indoors in pasteurized potting soil or in sterile peat pellets. Buy seeds that have been treated with a fungicide, or treat seeds with captan or thiram. Spray seedlings weekly with captan, ferbam or zineb. Avoid crowding, overwatering and overfertilizing seedlings.
CLUBROOT	Roots become a mass of club-shaped swellings. Cabbages and related vegetables may not form heads but remain stunted and yellowish. Plants may wilt on hot dry days, recovering at night. Outer leaves may yellow and drop. SUSCEPTIBLE PLANTS: BROCCOLI, BRUSSELS SPROUTS, CABBAGE, CAULIFLOWER, COLLARDS, RADISH, TURNIP	Plant in alkaline soil where these crops have not been grown for at least three years. Pull up and burn diseased plants.

Picture credits

The sources for the illustrations in this book are shown below. Credits from left to right are separated by semicolons, from top to bottom by dashes. Cover—Richard Meek. 4 —Keith Martin courtesy James Underwood Crockett; Richard Crist. 6—Courtesy Sy Seidman. 11,12,13—Richard Meek. 16—Courtesy Sy Seidman. 19,22,24,26,28,29,30,31— Drawings by Vincent Lewis. 33,34,35—Brian Seed. 36 —Drawing by Adolph E. Brotman. 38—Brian Seed except bottom center Lynn Pelham from Rapho-Guillumette. 39 —Lynn Pelham from Rapho-Guillumette. 40,41—Brian Seed. 42—Richard Jeffery. 45—Drawing by Vincent Lewis. 46—Brian Seed. 47—Brian Seed except top right Bettmann Archive. 48—Brian Seed except top right Bettmann Archive. 50 through 55—Drawings by Vincent Lewis. 56—Courtesy of Hunt Institute, Pittsburgh, Pa. 59—Drawing by Vincent Lewis. 65 through 77—Enrico Ferorelli. 78 through 147 —Encyclopedia illustrations by Richard Crist. 148,149— Maps by Adolph E. Brotman. 151,152,154—Drawings by Davis Meltzer. 153—Drawings by Davis Meltzer and Rebecca Merrilees.

Acknowledgments

For their help in the preparation of this book, the editors wish to thank the following: Dr. W. P. Bitters, Professor of Horticulture, University of California at Riverside; Alvin C. Blake, Assistant Professor, The University of Tennessee Institute of Agriculture, Knoxville, Tenn.; Dr. Robert Carlson, Horticultural Department, Michigan State University, East Lansing, Mich.; Frederick Corey, International Apple Institute, Washington, D.C.; Mr. and Mrs. William B. Crane Jr., Solano Beach, Calif.; Mrs. Edith Crockett, Senior Librarian, The Horticultural Society of New York, Inc., New York City; H. R. Denny, Don Perrine and Paul Stark Jr., Stark Brothers Nursery, Louisiana, Mo.; Dr. James Dewey and Dr. Warren Johnson, New York State College of Agriculture and Life Sciences at Cornell University, Ithaca, N.Y.; Miss Frances Ferguson, Weston, Conn.; Miss Marie Giasi, Librarian, Brooklyn Botanic Garden, Brooklyn, N.Y.; Mrs. Benjamin McFarland Hines, Greenfield Hill, Conn.; Mrs. Iva Inman, Sweeney, Krist and Dimm, Portland, Ore.; Richard A. Jaynes, Associate Geneticist, The Connecticut Agricultural Experiment Station, New Haven, Conn.; Mr. and Mrs. Jack E. Larson, Stafford Springs, Conn.; Charles Magoon, United Fresh Fruit and Vegetable Association, Washington, D.C.; William G. Moore, Production Manager, Jiffy Products of America, West Chicago, Ill.; Mrs. Eloise Ray, Westport, Conn.; Mrs. Murray Sargent Jr., Westport, Conn.; Donald Scheer, All-America Selections, Kingston, R.I.; Mrs. Sonia Wedge, Reference Librarian, New York Botanical Garden Library, Bronx Park, N.Y.

Bibliography

Abraham, George, *The Green Thumb Book of Fruit and Vegetable Gardening*. Prentice-Hall, Inc., 1970.

Anderson, Frederick O., *How to Grow Herbs for Gourmet Cooking*. Hawthorn Books, Inc., 1967.

Blair, Edna, *The Food Garden*. The New American Library, 1972.

Brooklyn Botanic Garden, *Fruit Trees and Shrubs,* Brooklyn Botanic Garden, 1971.

Brooks, Reid M., and Claron O. Hesse, *Western Fruit Gardening*. University of California Press, 1953.

Campbell, Mary Mason, *Betty Crocker's Kitchen Gardens*. Universal Publishing, Inc., 1971.

Childers, Norman Franklin, *Modern Fruit Science*. Horticultural Publications, 1961.

Clarkson, Rosetta E., *Herbs: Their Culture and Uses*. The Macmillan Company, 1942.

Foster, Gertrude B., *Herbs for Every Garden*. E. P. Dutton and Company, Inc., 1966.

Fox, Helen Morgenthau, *Gardening with Herbs for Flavor and Fragrance*. Sterling Publishing Company, Inc., 1970.

Jaynes, Richard A., *Handbook of North American Nut Trees*. The Northern Nut Growers Association, 1969.

Kains, M. G., *Culinary Herbs*. Orange Judd Company, 1912.

Kraft, Ken and Pat, *Fruits for the Home Garden*. William Morrow and Company, Inc., 1968.

Pellegrini, Angelo M., *The Food-Lover's Garden*. Alfred A. Knopf, 1970.

Rockwell, F. F., *10,000 Garden Questions Answered by 20 Experts*. Doubleday and Company, Inc., 1959.

Rodale, J. I., and Staff, *Encyclopedia of Organic Gardening*. Rodale Books, Inc., 1959.

The Royal Horticultural Society, *The Fruit Garden Displayed*. The Royal Horticultural Society, 1968.

The Royal Horticultural Society, *The Vegetable Garden Displayed*. The Royal Horticultural Society, 1961.

Simmons, Adelma Grenier, *Herbs to Grow Indoors for Flavor, for Fragrance, for Fun*. Hawthorn Books, Inc., 1969.

Simmons, Adelma Grenier, *The Illustrated Herbal Handbook*. Hawthorn Books, Inc., 1972.

Sunset Books, *How to Grow Herbs*. Lane Books, 1972.

Sunset Books, *Vegetable Gardening*. Lane Books, 1961.

Tukey, Harold B., *Dwarfed Fruit Trees*. The Macmillan Company, 1964.

Webster, Helen Noyes, *Herbs—How to Grow Them and How to Use Them*. Charles T. Branford Company, 1942.

Wescott, Cynthia, *The Gardener's Bug Book*. Doubleday and Company, Inc., 1964.

Index